Refurbishing Occupied Buildings:

The Management of Risk under the CDM Regulations

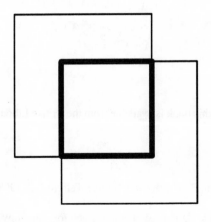

Bev Nutt
Peter McLennan
Roger Walters

Ͳ Thomas Telford

Published by Thomas Telford Publishing, Thomas Telford Ltd, 1 Heron Quay, London E14 4JD.

URL: http://www.t-telford.co.uk

Distributors for Thomas Telford books are
USA: ASCE Press, 1801 Alexander Bell Drive, Reston, VA 20191-4400, USA
Japan: Maruzen Co. Ltd, Book Department, 3–10 Nihonbashi 2-chome, Chuo-ku, Tokyo 103
Australia: DA Books and Journals, 648 Whitehorse Road, Mitcham 3132, Victoria

First published 1998

A catalogue record for this book is available from the British Library

ISBN: 0 7277 2732 X

This book is published on the understanding that the authors are solely responsible for the statements made and opinions expressed in it and that its publication does not necessarily imply that such statements and/or opinions are or reflect the views or opinions of the publishers. All reasonable effort has been made by the authors to ensure that the statements made and opinions expressed are safe and secure. No liability or responsibility can be accepted for any loss incurred in any way whatsoever by any person relying on the information contained herein.

Printed and bound in Great Britain by Bookcraft (Bath) Ltd

Foreword

The Construction (Design and Management) Regulations 1994 (CDM) present a challenge to all involved in the construction industry. CDM provides a framework for the effective management of health and safety throughout the lifecycle of a structure, from its initial inception, through construction and into subsequent maintenance, refurbishment and eventual demolition.

CDM requires a change in culture throughout the industry. Clients and designers now have to play a key part in generating the conditions in which construction work can be undertaken in safety and without generating risks to health. This book emphasises that to do this effectively managers, designers and contractors must all work together from the earliest stages of a project. The involvement and integration of all those who have knowledge of the work and the implications of that work is a central theme.

This book also focuses on the management of the risks associated with building refurbishment. These range from health effects to safety risks, affecting both those involved in construction work and those occupying the building whilst the work proceeds. How these risks can be evaluated and treated proportionately is discussed. Doing this effectively will ensure that consideration is given to the key risks and effort is not diverted to matters which will produce no health and safety benefit.

The construction industry is still learning how to apply CDM efficiently and effectively. This book contributes to this learning process by examining strategies for decision taking which can be applied to refurbishment works. The framework set out in the book should assist all those involved in refurbishment projects to understand their role under CDM and how to discharge it effectively.

Sandra Caldwell, HSE Chief Inspector Construction

Acknowledgements

Many have contributed to the production of this Guide, which is the result of research undertaken as part of the LINK Programme on Construction Maintenance and Refurbishment (CMR). We wish first to acknowledge the sponsors of this research, the Department of Environment, Transport and the Regions (DETR), the Engineering and Physical Sciences Research Council (EPSRC), and the ten industrial partners who collaborated in the work. The research team included Professor Bev Nutt, Peter McLennan and Dr. Charles Egbu from University College London, and Dr Roger Walters and Mike Lewis from Bickerdike Allen Partners. Professor Patrick O'Sullivan (UCL) and Professor Anderson (Bickerdike Allen Partners) acted as senior advisors to the project.

The ten industrial partners that collaborated in the research were: Bickerdike Allen (CDM) Ltd., Oscar Faber Consulting Engineers Ltd., Galliford plc, Stepnell Ltd., Heatherwood & Wrexham NHS Trust, Oxford Radcliffe Hospital NHS Trust, Worthing & Southlands Hospital, Norwich Union Assurance, Prudential Portfolio Management and Donaldsons Property Management. Additional expert advice was provided at no cost to the project by: B Holt (Willis Carroon), J Barber (Kings College, London), R Phillips (UMIST), D MacEwen (Shell International Petroleum Company Ltd), T Hetherington and S Peckitt (Health and Safety Executive) and the Craigavon Hospital Trust.

The research was reviewed and progressed by a Management Steering Committee comprising: P Pullar-Strecker (CMR LINK Committee Co-ordinator), R Kinnear and C Barber (CMR LINK Committee), N Jarrett (DETR), P Bates and J Williams (EPSRC), R Walters (Bickerdike Allen Ltd), S Robinson and P Benson (Oscar Faber Consulting Engineers Ltd), M Wakeford (Stepnell Ltd), D Chisnall (Heatherwood & Wrexham NHS Trust), M

Burrows (Oxford Radcliffe Hospital NHS Trust), B Banks (Worthing & Southlands Hospital), R Gowlett (Norwich Union Assurance), B Peal and E Wilson (Prudential Portfolio Management), J Bremner and S Collard (Donaldsons Property Management).

The significant contributions made by Steering Committee members and the external experts and advisors, are warmly acknowledged, as is the help of Kristina Andersen, Eva Culleton-Oltay and Joanna Saxon in producing this Guide. The project team is particularly grateful for the time and generous support given by the industrial partners throughout the project. Our thanks go to all.

September 1998   Bev Nutt
                                                  Peter McLennan
                                                  Roger Walters

Contents

Figures

Tables

1 Refurbishment risks in occupied buildings

1.1 Introduction

The principal aim of this *Guide* is to help to reduce the risks to human health and safety when buildings are refurbished while in occupation. In particular, the *Guide* seeks to assist all who are involved in refurbishment works to meet their obligations under both the Health and Safety at Work etc. Act 1974 (HSWA) and the Construction (Design and Management) Regulations 1994 (CDM).[1]

Refurbishment work represents some 42% of UK construction activity overall and it is generally considered to be more hazardous than new build construction.[2] Until such time as reliable data of refurbishment accidents becomes available, it can be assumed that at least half of all construction accidents probably occur during refurbishment projects.[3] The practical aim of the *Guide* is to provide a more secure framework for decisions and procedures to reduce these accidents and injury, not only while refurbishment work is being undertaken, but also during the subsequent periods of building use, re-use, refitting and modification, and when further refurbishments and adaptations are undertaken, until the time of demolition.

---

[1] Legislation drafted in response to European Council Directive 92/57/EEC, The Temporary or Mobile Construction Sites Directive, 24 June 1992.

[2] Egbu, C (1997) Refurbishment management: challenges and opportunities. *Building Research and Information.* vol 25, no 6, pp 338-347, and CIRIA (1994) *A guide to the management of building refurbishment.* (Report 133) UK: Construction Industry Research and Information Association.

[3] The Health and Safety Executive RIDDOR database does not classify reportable accidents into new build and refurbishment categories.

1.2 The CDM Regulations

The CDM Regulations 1994, provide a real opportunity to reduce the level of accidents arising from construction activity. The general purpose of the Regulations is to increase awareness of health and safety issues throughout the life of a building, from project inception, briefing, design, construction and commissioning, through to cleaning, maintenance, repair and refurbishment. In essence the legislation views a building as a product with a particular life span during which 'construction' activities take place and health and safety issues arise. The key participants in this life cycle are identified as client, planning supervisor, designers, principal contractor and contractors. The CDM Regulations provide the means for establishing working co-operation between these participants to help eliminate, reduce and control the hazards and risks that can affect the health and safety both of construction workers and building occupants. The specific legal obligations under the CDM Regulations, and the procedures, documentation and actions that they require, are detailed comprehensively in a number of publications.[4] They are summarised in a simplified form in this *Guide*, but are not repeated.

1.3 Key points of change

The CDM Regulations have created new legal responsibilities for the avoidance, reduction and control of health and safety risks, that affect both clients and building professionals alike. This legislation seeks to avoid, reduce and control health and safety risks within the construction industry by : [5]

---

[4] Health and Safety Commission (1995) *Managing construction for health and safety: Construction (Design and Management) Regulations 1994 Approved Code of Practice.* London: Health and Safety Executive, and Joyce, R (1995) *The CDM Regulations explained.* UK: Thomas Telford Services Ltd.

[5] Health and Safety Commission (1992) *Management of health and safety at work: Approved Code of Practice.* London: HMSO.

- co-ordinating the responsibilities of clients and building professionals through the use of two new documents: the Health and Safety Plan, and the Health and Safety File
- defining two new roles, those of the Planning Supervisor and the Principal Contractor
- introducing new statutory obligations and duties in relation to buildability and information support [6]
- requiring that formal risk assessments be undertaken in all notifiable projects

This *Guide* seeks to clarify the key issues raised by each of these changes and to provide practical assistance to those involved in the refurbishment of premises while in occupation. It specifically aims to:

- extend the existing CDMR guidance material, which concentrates on new-build construction activity, to cover the particular problems and risks associated with the refurbishment of occupied buildings
- provide a summary account of the requirements for health and safety in relation to the refurbishment of occupied buildings
- establish a practical organisational framework for co-ordination and communication, to support decisions and procedures for a more systematic approach to the management of health and safety risks during the refurbishment of occupied premises
- help to apply generic risk management procedures to the design, construction and management responsibilities of those involved in the refurbishment process
- support refurbishment decisions with an understanding of the available evidence concerning construction worker safety, accident frequency and the accident severity associated with various types of site activity

---

[6] see for example the explanation of these two requirements given in Barber, J (1997) "Potential side-effects of the CDM Regulations." *Construction Law Journal* vol 13, no 2, pp 26-37.

- summarise the practical preventative and protective measures that need to be co-ordinated throughout a refurbishment project

1.4 Areas of compound risk

Refurbishment projects involve more uncertainty and risk than new build construction.[7] Risks are increased further when an organisation needs to remain in a building or site while refurbishment is being undertaken. Under the existing health and safety legislation all employers have a duty of care for their employees. This duty of care does not discriminate between those inhabiting a building or those constructing a building. However, while the potential health and safety risks to occupants and the construction employees are very different, they coincide both in location and in time, when occupied buildings are refurbished. The area of compound risk, shown diagrammatically in Figure 1, can have detrimental affect and give rise to risks to:

- the **core business** of an organisation, through interruption or disruption of its work operations
- the health and safety of the **occupants**, both staff and the public
- the refurbishment **contract**, both its timing and its price
- the health and safety of construction **workers**,[8] and all associated suppliers and contractors when on site

Typically, the level of compound risk will be a consequence of the physical characteristics of the existing building, its site and access, the nature of the organisation's operations, and the type of the construction and refurbishment activities that are to be undertaken. This *Guide* focuses on these areas of compound risks, both to the

---

[7] Egbu, C (1995) Perceived degree of difficulty of management tasks in construction refurbishment work. *Building Research and Information.* vol 23, no 6, pp 340-344 and Health and Safety Executive (1988) *Blackspot construction.* London: HMSO.

[8] The CDM Regulations broadens the construction employment group to include those that maintain and clean buildings.

building occupant and to the construction worker. The *Guide* will help to identify :

- areas of significant health and safety risks in construction projects in general and in occupied refurbishment projects in particular.
- the circumstances that can create compound risk
- the interaction between refurbishment risks and building occupancy risks
- the implications for project planning and management.

Figure 1   Refurbishment risks in occupied buildings

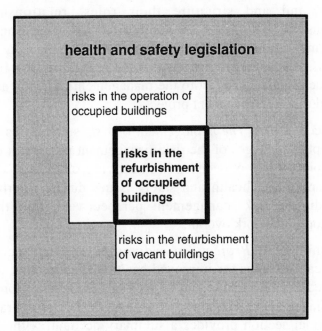

1.5 Structure of the *Guide*

This *Guide* is arranged in three parts concerning:

- the organisation of health and safety responsibilities
- the assessment of health and safety risks
- the project co-ordination of health and safety measures

These three parts have been written on a 'stand alone' basis. Each can be used independently of the other. The three parts of the guide, taken together, provide both generic and project specific guidance concerning the refurbishment risks in occupied buildings, supported by the findings of construction safety research, the available construction health and safety accident data, and published best practice procedures.

The first part, A *decision framework for CDM*, identifies the major areas for decision, the main participants that are involved, and the key interfaces for communication and co-ordination under the CDM Regulations. This framework has been developed to help those that are involved in refurbishment projects to clarify, understand and structure their roles, relationships and responsibilities, one with another, in the development and implementation of safety strategies, safety plans and safety actions. It is intended that the decision framework will be used to provide practical support to the management of health and safety risks during refurbishment projects.

The second part, *The management of risk*, summarises the main concepts and stages of the risk management process. It describes the typical stages of decision in identifying, assessing, controlling and monitoring health and safety issues during refurbishment, providing a risk management perspective within the CDM decision framework overall.

The third part, *Project co-ordination*, sets out the range of practical measures for improved health and safety that should be adopted by the various participants in the Design, Construction and Management components of the CDM Decision Framework. This final section provides a summary account, with check list support, of the practical application of the Decision Framework to meet the specific circumstances of a given project. Taken together, these three sections provide a comprehensive basis for understanding and implementing the CDM Regulations when occupied buildings are to be refurbished.

2 A decision framework for CDM

2.1 Project participants

All of those that contribute to the planning, design, management, construction and commissioning of a refurbishment project need to be aware of the potential health and safety risks that might arise. When occupied buildings are refurbished, the risks are potentially greater. The health and safety of employees, visitors, customers and the public can also be put at risk. The refurbishing of buildings while in use can have detrimental effect on an organisation's business. The client needs to be fully aware of any risks that a proposed design, and the construction activity that it implies, might hold for core business activity. The risk of accidents on site are likely to be increased further if there are tight financial constraints to the project and if unreasonable time pressures are imposed by project management and poor design. So in occupied buildings, the safety of all of those that are involved in, or might be affected by, the refurbishment process must be considered from the start.

The potential risks of refurbishment may be increased or decreased in three ways; through management procedures and actions, as a result of design concepts, decisions and specifications, and through construction methods and activities on site. Each of these three areas, to which the CDM Legislation relates, involves two types of participant:

- Management   -   the Client organisation
                -   the Facilities Management Team
- Design   -   the Design Team
                -   the Planning Supervisor
- Construction -   the Principal Contractor
                -   the Contractors

HSC,[1] HSE,[2] and CIRIA[3] publications provide guidance concerning participant roles under the CDM Regulations and they are described in detail in section 4.1 of this *Guide*. The roles of five of the participants listed above are specifically defined in the CDM Regulations. The exception is the role of the facilities management team. The Regulations are not directed to the health and safety problems of refurbishment specifically, nor do they address the additional risks that occupied buildings entail. No reference is made in CDM to the potential contribution of the facilities management team. The facilities manager's detailed knowledge of a building's characteristics, the organisation's operations and support services, provides a valuable source of additional expertise and information for health and safety risk reduction. In effect, the facilities management team is often ideally placed to act as the 'expert client'.

2.2 Levels of decision

In each area of involvement (management, design, and construction) participants will have opportunities to avoid, reduce and control risks at three basic levels of intervention:

- **the strategic level**: dealing with strategic issues and safety strategy
- **the operational level**: dealing with operational issues and safety plans
- **the tactical level**: dealing with tactical issues and safety actions.

---

[1] Health and Safety Commission (1992) *Management of health and safety at work: Approved code of practice.* London: HMSO.

[2] see HSE information sheets: no 39, The role of the client; no 40, The role of the planning supervisor; and no 41, The role of the designer. Also, Construction Industry Advisory Committee (1995) *Designing for health and safety in construction: A guide for designers on the Construction (Design and Management) Regulations 1994.* London: HMSO.

[3] CIRIA (1997) *CDM Regulations-Practical guidance for clients and clients' agents,* (Report 172) CIRIA: London and CIRIA (1998) *CDM Regulations-Practical Guidance for Planning Supervisors,* (Report 173). CIRIA: London.

A basic decision framework, linking the six participant types and the three decision levels,[4] is shown in Figure 2. This simple framework places the participants specified in the CDM Regulations (client, planning supervisor, designer, principal contractor, contractor) along the horizontal axis and the hierarchy of decision issues (strategic, operational, tactical) on the vertical axis. Since health and safety legislation generally, and the refurbishment of occupied buildings in particular, must address the problems of buildings-in-use, the facilities management team is included alongside the client.

Figure 2 Levels of decision

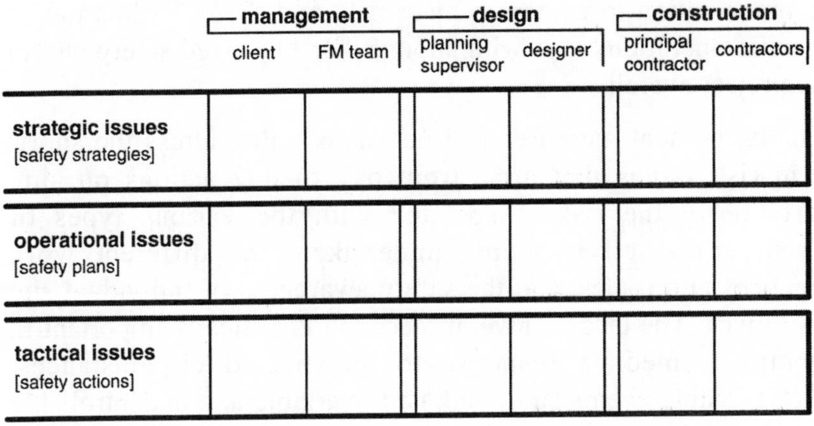

| | management | | design | | construction | |
	client	FM team	planning supervisor	designer	principal contractor	contractors
strategic issues [safety strategies]						
operational issues [safety plans]						
tactical issues [safety actions]						

The strategic level of decision is primarily concerned with long-term risk issues, extending throughout the life of the project, from briefing to practical completion. Strategic level decisions will tend to be of fundamental importance with project wide significance, focusing on those issues that can have major impact on the success of the project as a whole. They will also be directed at the long-term risks beyond the life of the project, addressing the safety

---

4   The distinction between strategic, operational and tactical levels of decision used in this *Guide* relates specifically to CDM issues. It is at variance with definitions used in some standard management texts.

implications of today's decisions for those who will service and maintain the building after refurbishment, the health and safety of the next generation of occupants and the safety of contractors who may undertake further construction work in the future.

In contrast, the operational level of decision shown in Figure 2, relates to the medium-term risk issues associated with the various stages of work and the specific site processes and operations that are to be undertaken. In refurbishment projects the safety implications of construction operations should be anticipated in terms of time, sequence and method. The planning of all site operations can have significant impact on site safety, to a far greater extent than in new-build projects. Decisions here, will focus on the detailed issues of organisation, management and communication to contain project risk, and the co-ordination of specific operations, one with another, for improved safety within the project overall.

Finally, tactical measures and decisions will address the short-term risk issues that arise from day to day actions on site, particularly the risks associated with the various types of construction activity being undertaken, the different work practices and trades, and the safety awareness of individual site operatives. The tactical level of decision is critically important in making immediate response to unexpected circumstances, unforeseeable events and to adjust to variable and uncontrollable local conditions under which work may need to be conducted.

Strategic, operational and tactical safety measures need to support and reinforce each other within the overall framework for the management of the project. Strategic risks need to be avoided and reduced through the implementation of secure 'Safety Strategies'. Operational risks need to be reduced and controlled through 'Safety Plans' covering all major operations. Tactical safety issues need to be faced through the introduction of good working practices, risk adverse codes of conduct and the adoption of agreed site 'Safety Actions'. Finally, each of the three levels of decision commonly require that contingency arrangements be put in place so that unpredictable situations and risks can be contained, should and when they arise.

2.3 Project communication

Effective communication between all project participants, at strategic, operational and tactical levels, is a prerequisite for the successful containment of risk. Five primary interfaces for project communication are identified in Figure 3. The first three relate to the key areas of communication and co-ordination; Management to Design [I/1], Design to Construction [I/2] and Construction to Management [I/3]. The remaining two interfaces are critical in linking the strategic, operational and tactical levels of decision.

The interface between the Strategic and Operational safety issues [I/4] relates the Safety Strategies adopted at the beginning of a project, to the Safety Plans covering the specific operations that are to be undertaken within these strategies. Finally, the fifth interface between the Operational and Tactical safety issues [I/5] relates to the lines of communication between those responsible for safe operational planning and those with responsibility for safe behaviour and actions on site.

Figure 3   The five project interfaces

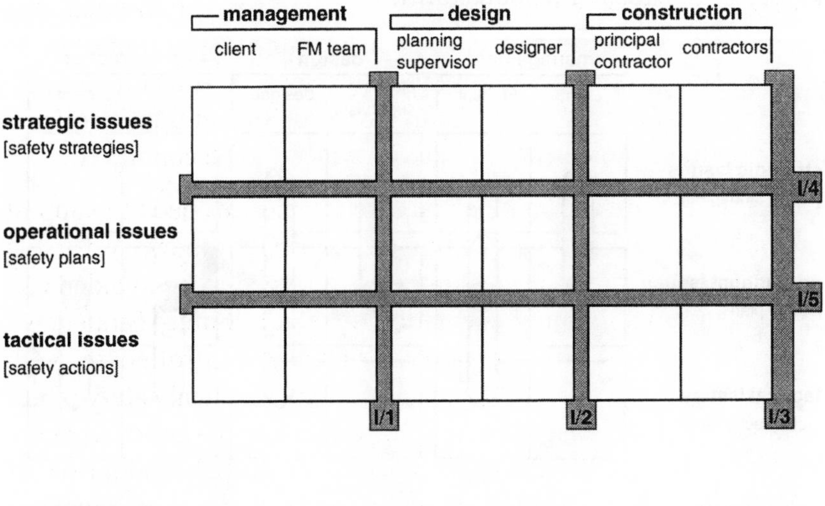

Refurbishment projects often entail hidden risks due to partial or unreliable information concerning the 'as found' conditions, materials and structure of a building. The practical difficulties of on-site working cannot therefore, always be anticipated fully in advance. This uncertainty, due to the hidden risks of refurbishment, highlights the need for clear procedures and effective communication across each of the five primary interfaces, shown in Figure 3, in the interest of both the client organisation and the contracting firms alike.

2.4 Areas of responsibility

Each type of project participant has different areas of responsibility under the CDM Regulations. The decision domains of the various participants are indicated in Figure 4. Here, the vertical arrows show the typical range of participant responsibility for strategic, operational and tactical health and safety issues. Client responsibilities are primarily focused on strategic issues, as are those of the designer. While their strategic concerns should be expected to inform operational and tactical safety issues, neither the client nor the designer will normally be responsible for work activities on site.

Figure 4   Areas of responsibilities

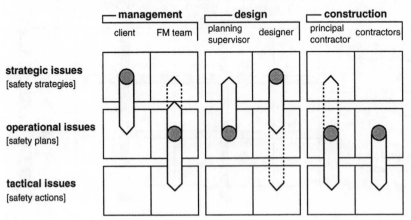

The responsibility for health and safety co-ordination rests with the planning supervisor and the principal contractor, working together to bring all significant operational and tactical risks under control. In refurbishing occupied premises, the expertise of the facilities management team can provide a 'third arm' of support, bringing operational and tactical safety issues to the attention of the contractor. It is the planning supervisor, principal contractor and FM team therefore, who hold a pivotal role in refurbishment projects, focusing on operational issues, but relating these upwards to inform long-term safety concerns at the strategic level, and downwards to help tackle day to day safety issues of work on site.

2.5 Levels of co-ordination

Three primary levels of project co-ordination are implied in the CDM Regulations. These are shown in Figure 5. The [C/1] level of co-ordination represents the strategic integration required by the client, the planning supervisor and the design team early in the project. The appointment of the principal contractor is preferable at this stage of co-ordination, since the issues of buildability can significantly affect the viability of a project, and the health and safety problems that might arise. At the very least the safety issues for management and design must be co-ordinated at the strategic level in order to clarify and specify the range of safety issues that will need to be resolved at the operational level.

The [C/2] level of co-ordination shown in Figure 5, involves the planning of all operations and activities prior to the commencement of work on site. This stage involves the development of the Health and Safety Plan as the primary co-ordinating document, identifying all significant health and safety risks. This plan should fully anticipate the tactical safety issues at the third [C/3] level of co-ordination. This level considers the detailed arrangements for day to day working practices and activities on the site. Two types of activity must be considered in parallel, those relating to the construction processes and those relating to the work of the client organisation. Areas of significant risk require pre-planning at an early stage. Here the importance of client liaison in reducing health and safety risks, should not be underestimated.

Figure 5   Levels of co-ordination

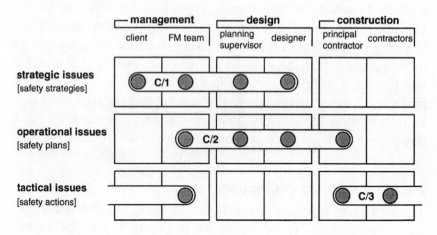

2.6   The decision framework

The CDM decision framework, as set out in this Chapter, and summarised in Figure 6, provides a generic basis for the systematic consideration, review and co-ordination of health and safety measures under the CDM Regulations, at strategic, operational and tactical levels of intervention.

Figure 6   CDM decision framework

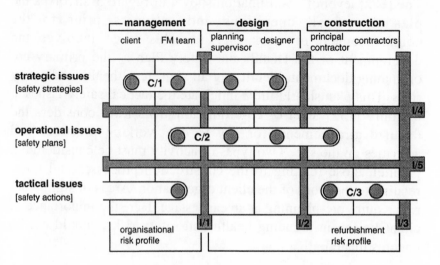

This framework covers the responsibilities of six types of participant in the refurbishment process, at the five primary interfaces for co-ordination and communication that are required:

- management to design
- design to construction
- management to construction
- strategic level to operational level
- operational level to tactical level.

The framework can be used as an organisational tool with which to examine the health and safety risk profile presented by any given refurbishment project. It can also be used to simplify, interpret and structure the extensive but unmanageable health and safety checklists that are available (see Appendix A1). The framework provides a basis for considering and agreeing practical working arrangements in the light of the relative complexity or simplicity of a project, identifying project requirements and priorities, unusual circumstances and any unique risks that might be involved. The decision framework can be used to assist in :

- **agreeing responsibilities**: considering professional and project responsibilities for each of the six participant types in the context of the scale and complexity of the project in-hand, formally agreeing duties at strategic, operational and tactical levels
- **defining relationships**: clarifying roles and arrangements for collaboration, co-ordination and communication between participants and their areas of decision, across each of the five key interfaces
- **identifying risks**: reviewing the risks and hazards that might arise within the context of a specific project, both for the organisation's activities on the one hand, and for refurbishment activities on the other
- **managing risk**: establishing effective arrangements for safety management, response and control for the reduction of health and safety risks, with clear procedures linking safety strategies to safety plans, and linking safety plans to safety actions
- **co-ordinating documentation**: agreeing the form and extent of project records concerning safety strategies, safety plans and

safety actions, the latter two being mandated by the CDM Regulations through the required Health and Safety Plan and the Health and Safety File documentation.

Finally the Decision Framework provides the means for analysing and profiling the areas of compounded risks that are associated with a specific refurbishment project. The two basic risk profiles, the organisational profile and the refurbishment project profile, are shown at the bottom of Figure 6. The client and the FM team will normally be responsible for defining the organisational risk profile. The groups examining and specifying the refurbishment risk profile will include designers, facility managers, principal contractors, and in some projects sub contractors may also be involved. This risk identification, and the risk management process of which it is part, will be considered next.

3 The management of risk

3.1 The risk management process

The basic components of a typical risk management process are shown in Figure 7. In order to reduce risk exposure overall, this process needs to operate at the three levels described in Chapter 2; strategic, operational and tactical. The review, feedback and auditing stages shown in Figure 7, help to reduce risks and improve management performance in the longer term. The other five stages are undertaken within the timescale of a given project.

Figure 7   The risk management process

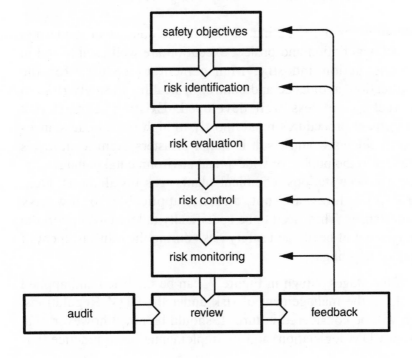

The basic purposes of the risk management processes are to:

- clarify objectives
  what outcomes are to be avoided?
  what types of risk are to be considered?
- identify areas of risk
  what risk data is available?
  what are the potential risks?
  how likely are they to occur?
- evaluate the significance of risk
  how can degrees of risk be assessed?
  how serious is their potential impact?
- manage and control risk
  which risks can be eliminated?
  which risks can be transfered to others?
  how can the remaining risks be reduced?
  how can the residual risks be controlled?
- monitor performance
  are risks being effectively contained?
  is the management of risk improving over time?

Procedures for the handling of risk in the context of tendering, contract negotiation and project execution are well established in the construction industry. Management procedures for the identification, avoidance and control of health and safety risks in construction are less well developed. Existing 'expert' risk management procedures relate mostly to high risk circumstances where accidents would result in major disasters, significant levels of danger to the public, or widespread environmental damage (e.g. nuclear power stations, chemical plants, gas installations, etc.). These procedures are not directly applicable to low risk circumstances. They need to be specifically tailored to support the management of health and safety risks during the refurbishment of occupied buildings.

All of the stages shown in Figure 7, can be modified and applied to help in the management of risk under the CDM Regulations. Each will be considered in turn. It should be noted however, that current CDM legislation, and its implementation in practice, has tended to place very high priority to the risk assessment stage, rather than adopting a balanced approach to all stages of the risk management process, of which the assessment stage is part. The exposure to risk typically relates to a combination of hazards and risks including:

- **safety**: the risks of injury or death to staff, customers and the public, the risks to personal security, health and well-being
- **business**: the risks of disruption to production, operations, sales and services, the loss of financial and human assets and intellectual property
- **assets**: the risks to investments, property and equipment
- **environment**: the risks to the local and global environment and to public health generally

The CDM Regulations are directed only to the first of these four areas of risk exposure, that of health and safety. Furthermore, the Regulations are narrowly focused. They relate only to the health and safety of employees while at work, and to other persons that are directly affected by the work undertakings. However, in setting up health and safety risk management procedures for refurbishment projects for buildings while in use, it is wise to consider the inclusion of the other three areas of risk, particularly those relating to the continuity of business operations.

3.2 Safety objectives

The fundamental objective of the Health and Safety Legislation is the avoidance, reduction, and control of health and safety risks to those at work. It is assumed that the existing statutory provisions and obligations for health and safety are already understood by those reading this document. They are typically administered by the facility management team of an organisation, or by a contractor's health and safety manager when refurbishment works are undertaken. Details are given in a number of health and safety handbooks and a list of references is included in Appendix A1. The CDM Regulations provide general guidance in this area, but the main source of advice is the Approved Code of Practice under the Management of Health and Safety at Work Regulations 1992.[1]

It is essential that clear objectives are established for the management of health and safety before reburbishment work is begun. Objectives must be unambiguous, specifying which health and safety risks are to be managed, defining all of those to whom the health and safety measures will apply, and the time periods for health and safety responsibility. In the case of refurbishment projects for occupied buildings, health and safety measures must be put in place to protect the health and safety of at least four constituences:

* all employees of the contractor and subcontractors while they are working on the site
* all employees of the client organisation while on site
* all other building occupants, for example, service providers, patients, visitors, customers, suppliers etc., during all times when they are on the site
* all other persons and members of the public who might be put at risk from the work being undertaken

All risks that might lead to death, 'major' injury, 'over-3-day' injury[2], or damage to health must be addressed and defined within a statement of objectives (detailed definitions of these categories of injuries are given in Appendix A2.1). The objectives may form a specific part of an organisation's health and safety policy, or they may be incorporated within the overall risk management policy for the organisation as a whole. The statement of objectives must include the management's policy for meeting health and safety obligations, with clear directions concerning the identification, evaluation and control of health and safety risks. In addition, the statement of objectives should include monitoring, reviewing and

---

[1]   see for example Health and Safety Commission (1992) *Management of health and safety at Work 1992 Approved Code of Practice*. London: HMSO, particularly Reg (3) page 2.

[2]   Over-3-day, together with death and major injury, is one of the three main categories of reportable injuries under the Reporting of Injuries, Diseases and Dangerous Occurrences Regulations 1995 (RIDDOR).

feedback arrangements for improvements to the management of health and safety over time.[3]

3.3   Risk identification

The hazards and risks associated with work activities need to be identified first. Hazards are defined as situations with the potential to cause harm or loss. Under the CDM Regulations it is both hazardous circumstances and hazardous activities that potentially threaten the health and safety of employees. So the identification of hazards and activities is an essential part of analysing the risks that these represent to health and safety in the workplace. Risk is a function of three factors; the likelihood of actual harm resulting from a hazard, the potential severity of the consequence should it occur, and the number of people that might be exposed to the harm. There are a number of well established methods by which risks may be quantified and hazard indices developed, if reliable data is available. Risk is normally measured in terms of the probability of a worker being injured in any given year, while carrying out his or her work, by the severity of injury suffered.[4] Accident frequency is often used as the prime indicator of risk overall.

---

3   These concepts are more fully detailed in: Dickson GCA (1995) *Corporate risk management.* London: Witherby; Confederation of British Industry (CBI) (1991) *Developing a safety culture: Business for safety* . UK: CBI; and Health and Safety Executive (1991) *Successful health and safety management.* London: HMSO.

4   The HSE expresses risk as "the likelihood that the harm from a particular hazard is realised." (para 5 (b)) Health and Safety Commission (1992) *Management of Health and Safety at Work 1992: Approved Code of Practice.* London: HMSO. The Royal Society expresses risk as: "the probability that a specified undesirable event will occur in a specified period or as a result of a specified situation." Royal Society Study Group (1992) *Risk: Analysis, perception and management.* London: Royal Society. Grimaldi and Simmonds (1984) *Safety management.* USA: RD Irwin, express risk as: "the assumed effect of an uncontrolled hazard, appraised in terms of the probability it will happen, the maximum severity of any injuries or damages, and the public's sensitivity to the occurrence." pp 181.

The identification of hazards and activities that can give rise to generic health and safety risks will rely on the available data concerning the frequency and severity of injuries that occured in the past. Accident data is limited and none of the available statistics distinguishes between injuries incurred during refurbishment work, new build or infrastructure construction. Furthermore, construction accident data is limited to formally reported injuries only. Unlike occupational health and safety data, 'near miss' accidents are not reported. As a result, it is especially difficult to quantify and evaluate the health and safety risks of refurbishment.

With no source of data specifically related to refurbishment accidents, three general sources of statistical data can be used to investigate the risks involved; the EU Safety Statistics,[5] the Registrar General's Occupation Mortality Data,[6] and the HSE Accident Statistics from the Reporting of Injuries, Diseases and Dangerous Occurrences Regulation (RIDDOR) safety data.[7] The EU data is limited to fatality statistics, aggregated on an industry by industry basis, by nation state. The RIDDOR database is the major source of data on construction injuries, detailing accidents and the activities being undertaken when accidents took place. This is therefore the primary source of statistical information to support risk identification, supplemented by the proportional mortality ratios (PMRs) from the Registrar General's data.

For the purposes of this *Guide*, the RIDDOR work accident data has been analysed to help to establish the chances of injury across the various activities of the construction industry, and to indicate priority areas for health and safety measures. A detailed analysis of 'major' and 'over-3-day' injuries has been undertaken for the period 1985–96, broken down by class of occupation, the activity being undertaken at the time of the accident, and by the type and

---

[5] European Health and Safety Agency, Bilboa, Spain.
[6] Office of Population and Census and Health and Safety Executive (1995) *Occupational health: decennial supplement.* F Drever (ed) London: HMSO.
[7] Health and Safety Executive, Statistical Services Unit, Bootle, UK.

severity of injury suffered. The results of this analysis indicated that:

- deaths from falls, accidental injuries and asbestos related illnesses are significantly higher in the 17 job categories associated with construction than in any other occupation.[8]
- the majority of work injuries are directly related to the class of occupation, type of activity and type of work process.
- of 18 occupational types 31.2% of all 'major' and 'over-3-day' injuries occurred in two occupational classes, those of labourer (19.7%) and carpenter/joiner (11.5%).[9] Contrary to expectations, scaffolders and ground workers had some of the lowest accident occurrences, 2.7% and 2.1% of all cases respectively.
- four activity types (out of 43) account for 50–80% of all serious accidents on site, these are finishing processes, transfer of materials, other processes, and ground works.
- the two dominant causes of serious injury were 'falls' (45–81%) and 'trip or slip' (32–58%) across the five most vulnerable occupation types. 'Falls' were the dominant cause of serious injury across the 10 most risky types of work process (30% - 79%).

Tables 1 and 2 summarise the current position, ranking those site operations, HSE process types and occupations that give rise to high accident rates, for 'major' and 'over-3-day' injury, respectively. These generic areas of risk will all require specific consideration during any project. Occupational health data provides mortality statistics for the long term health consequences of different occupations[10]. These statistics reinforce the HSE injury safety data, with proportional mortality ratios (PMRs) indicating that death from falls and accidental injuries are

---

8   this is consistent with data from other national safety and health bodies, Hinze, et al (1995) and Robinson et al (1995) review of data from the National Institute for Occupational Health and Safety (NIOSH) in the USA.
9   Culver et al (1993) report similar results in the USA.
10 For more complete explaination see the Occupational Health Decennial Supplement 1995.

significantly higher for construction workers than in all other occupations. Additional information of specific health related risks for 17 job group categories in construction are also included (see Table A23 in Appendix A2). Health hazards are dominated by increased PMRs due to asbestos and asbestos related diseases. This represents a significant health threat to all construction workers, endorsing the rigorous measures that have been put in place for the management of refurbishment work in asbestos environments.

Table 1   Major injury

site operations	process type	% reported accidents	occupations affected (rank order)
surface finish	finishing processes	22.7 %	carpenter/joiner electrician labourer
manual handling	transfer on site	9.9 %	labourer all managerial non-construction
other processes	other processes	8.9 %	labourer other occupations non-construction
groundworks	groundworks	8.3 %	labourer groundworker non-construction
structural erection	roofing	7.3 %	slater/roofer labourer carpenter/joiner

A more comprehensive summary of the analysis of work accident data is included in Appendix A2 at the end of this *Guide*, Tables A1 - A6 for 'major' injury and Tables A14 - A19 for 'over-3-day' injury. This provides generic reference material to those considering the actual hazards and specific risks associated with a particular project.

Table 2   Over-3-day injury

site operations	process type	% reported accidents	occupations affected (rank order)
surface finish	finishing processes	26.5 %	carpenter/joiner plumber/fitter labourer
other processes	other processes	9.9 %	non-construction labourer manual workers
manual handling	transfer on site	9.8 %	labourer carpenter/joiner non-construction
groundworks	surfacing (roads)	6.9 %	paviour labourer all managerial
delivery & site storage	load/unloading	6.3 %	labourer driver paviour

Risks identification for a specific project and its unique circumstances, will need to examine four principal types of hazard:

• the 'generic' hazards of refurbishment work

- the 'as existing' operational hazards arising from the normal working practices of the client organisation and building occupants
- hazards due to the 'as found' physical characteristics of the building, site and location
- the 'compounded' hazards that can arise due to the proximity of construction and operating activities, when refurbishment begins

Details of normal operational hazards can be obtained from an organisation's health and safety policy, their past records of occupational injuries and near misses, with day to day knowledge of health and safety problems provided by the facilities management team. The likelihood of 'compounded' hazards will differ with building type. For example, the potential hazards concerning patient health while hospital refurbishment is being undertaken, and the additional hazards to the public that may arise when retail premises or transport facilities are being refurbished, can be considerable.

There is a high degree of concensus that the risks of refurbishment are closely related to the nature of the existing structure, its form of construction, its materials and services.[11] By the time that most major refurbishments become necessary, reliable and comprehensive details of these physical characteristics are rarely available. Discrepancies between formal records and actual site conditions are commonplace. Existing documentation and records need to be systematically assessed, supplemented by inspections and surveys, to uncover all of the likely hazards that need to be considered. Secure data about the existing site and building, is usually the most critical area of uncertainty when selecting appropriate measures for the avoidance, reduction and control of refurbishment risks. The client is now obligated, under the CDM Regulations, to provide this information to the planning supervisor and principal contractor.

---

[11] Building Maintenance Information (1991) *Safe and effective building maintenance*. UK: RICS.

Within any set of identified hazards, health and safety risks will be affected by the nature of site and construction activities that are to be undertaken. Here the accident frequency and severity data suggests that two classes of activity should always be considered:

- activities that are inherently risky
- activities that are not risky per se, but do give rise to a high number of accidents

The perception that some activities such as demolition, excavation, steel erection, scaffolding and roofing, are inherently risky, is obvious. It is encouraging to note that accident data indicates that the level of injury actually sustained, while undertaking high risk activities, is on the low side. It seems that, when activities are considered to be dangerous, then greater care and control of health and safety risks is maintained. Naturally all high risk activities must continue to be identified and controlled, with priority given to the elimination and reduction of risk through the strategic and operational management measures discussed in Chapter 2.

Accident data suggests that most major injuries on site arise from the second class of activity, those ordinary activities that are not inherently risky. For example, the activities that give rise to some of the highest accident frequencies relate to 'material handling'. Some 300 of the 3000 serious accidents per annum arise while materials are being handled on site. It would appear from the available evidence, that health and safety practices directed specifically at these seemingly low risk activities, might lead to further reductions in the total number of accidents per year, benefiting the greater number of individuals overall. Improvements here are most likely to stem from a review of routine site procedures and work practices, targeted training and on-site briefing to improve risk awareness, all with reliance on the tactical management measures outined in Chapter 2.

3.4 Risk evaluation

Any evaluation of health and safety risks should, at the very least include:

- a comprehensive analysis of the hazards, processes and activities, identified as sources of significant risk
- the consideration of measures that could be established to control these areas of significant risk during a refurbishment project
- an examination of the ways in which significant risks may be managed during the remaining life of the facility after refurbishment.

Risk management decisions involve both the prediction of future hazards and risks, and arrangements with which to face the uncertainties associated with unforeseeable circumstances and chance events. The health and safety risks to employees need to be evaluated in relation to:

- foreseeable and unforeseeable risks
- significant and insignificant risks

Foreseeable risks are those that may be clearly identified within established working practices, under the existing legislation. The nature of refurbishment is such that some of the risks are likely to be unforeseeable due to key information about the building 'as found' being unavailable, not properly documented, or readily open to inspection. What is important to consider are the means by which any unforeseen risks are to be dealt with by planned contingency measures, and through effective communication and co-ordination arrangements throughout the project. Flexible response systems need to be put in place to help all participants to react to a changing awareness of hazards and risks as more information becomes available.

The CDM Regulations require that formal risk assessment be undertaken in areas of significant risk. The procedures for doing this arise out of earlier legislation, that of the EU Directive on Health and Safety, interpreted in the UK through the Approved Code of Practice under the Management of Health and Safety at Work Regulations, 1992 (MHSWR). This Code of Practice stresses the importance of assessing all 'significant' health and safety risks, and the emphasis on 'significant' risks is clearly implied in the 1994 CDM Regulations. No definition of 'significant' risk is included in this or any other health and safety legislation, however. The closest concept is that of 'tolerability', proposed by the Health and Safety Executive (HSE).[12] They have established a tolerability level for worker fatality at 10^3, that is not more than one fatality per 1,000 actively employed construction workers per year. 'Tolerability' thresholds of this kind help to indicate the level of accidents that might be considered to be 'significant'. Other potential sources, such as employers' liability insurers, consider this information to be commercially sensitive and not for general publication.[13]

Current HSE fatality data indicates a UK rate of 10^4 (one fatality for every 10,000 actively employed construction workers per year). So actual fatalities due to construction activity do not appear to be unreasonable in relation to the HSE's tolerability limit. While zero fatality must always be the prime health and safety objective, the current level of fatalities taken alone, would not require that a special risk assessment be undertaken. Deaths arising from construction activity may not be at intolerable levels, but the position in regard to non-fatal accidents, both those causing major injuries and 'over-3-day' injuries, is rather different. Current data here shows major incidents at an annual level of 10^3 (one major injury for every 1,000 actively employed construction workers per year). The data for 'over-3-day' incidents indicates an annual level of 10^2 (one 'over-3-day' injury for every 100 actively employed construction workers per year).

Figure 8 shows the range of incidents by severity; from fatal, major, 'over-3-day', minor injury, to non-injury accident. The five types of accident are shown in the form of an accident triangle that

---

12 The public inquiry into nuclear power plant safety established this tolerable level for fatalities, HSC (1988) *The tolerability of risk from nuclear power stations*. HMSO: London, Figure 6 pp 34.

13 Leopold and Leonard (1987) mention employers' liability ratings and give two examples, plasterers banded from 0.30%-0.90% of total wages compared to roofers banded at 3.0%-7.5% of total wages in a discussion of trade based premiums. These ratings are an insurers' indication of relative safety risk. Several commercial firms confirmed that these data are confidential, thus not available for public use.

indicates the approximate distribution of accidents by type;[14] few being fatal, most being minor, etc. Three thresholds for different accident severities have been quantified on the basis of the RIDDOR accident statistics. Since the RIDDOR data does not include minor injury and near miss accidents, the threshold between these two cannot be calibrated.[15] Estimates of accident severity are currently based on rather subjective criteria. Until such time as the health and safety statistics permit a more quantitative approach, the risk management concept of "As Low As Reasonably Possible", as recommended by the HSE,[16] should be applied.

Figure 8 The accident triangle

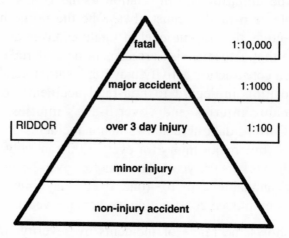

---

[14] Used in industrial safety as a method for illustrating the progressive effects of lower level incidents on the upper level incidents, typically as a series of ratios. Safety is controlled by monitoring the lowest levels which are viewed as 'near misses' for the next level up. see Heinrich et al (1980).

[15] Hubbard and Neil (1986) reported from a UK case study that they had 1 major accident for every 32 minor accidents. Typically, this data is not kept by contractors.

[16] HSE (1989) *Quantified Risk Assessment: Its input to decision making.* HSMO: London and Cox SJ and EF O'Sullivan (1992) *Building regulation and safety.* (BRE Report), UK: Construction Research Communications Ltd, Figure 4, pp 57.

The "As Low As Reasonably Possible" (ALARP) concept, is illustrated in Figure 9. This distinguishes between those risks that are intolerable and cannot be justified, from those that are tolerable from a practical viewpoint. The resulting ALARP region between the upper and lower thresholds, indicated in Figure 9, relates to risks that are not in themselves significant enough to warrant that a formal risk assessment be undertaken under the CDM Regulations. While practicable measures should always be undertaken to reduce tolerable risks still further, risks within the ALARP region do not in themselves constitute unreasonable levels of risk. Risks at this level can be handled informally in a routine manner, but the diligent organisation may none the less decide that formal risk management procedures should nevertheless be adopted.

Figure 9   The ALARP triangle

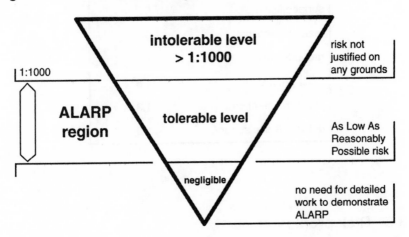

The HSE has positioned the threshold between tolerable and intolerable levels of risk at 1:1,000. This is in line with the RIDDOR major accident data. Accident frequencies above this 1:1,000 level must be considered as constituting 'significant' safety risks, which by definition are unacceptable. Here, formal risk assessments must be undertaken. The lower threshold, between tolerable and negligible levels of risk, is at present undefined and therefore needs to be determined on a project by project basis.

Table 3 shows the areas of significant risk that are currently above the ALARP 1:1,000 threshold. A detailed summary of this information by trades, construction activities and injury type, is provided in Appendix A2. Little data is available on the exact causes for these injuries. None is available in relation to refurbishment projects. However, this is the best information that is available to indicate the areas of significant safety risks during construction operations on site.

Table 3   Significant risks

site operations	incident ratio per 1,000 employees
surface finish	5.2
manual handling	3.5
groundworks	2.7
other processes	1.8
structural erection	1.3
delivery & site storage	1.3

3.5 Risk profiles

The possible combined effects from individual hazards and risks should be considered next. In the refurbishment of occupied buildings, the potential impact of compounded areas of risk must be investigated. Two risk profiles need to be examined; the profile of risks relating to the normal working operations of the organisation, and the profile of the risk associated with the refurbishment activities that are to be undertaken. A basic proforma for profiling the 'generic' and 'case specific' risks in each of these two areas, is illustrated in Figures 10 and 11 on pages 34 and 35. These list the generic operations of an organisation on the one hand, and the generic operations of refurbishment activity on the other. A number of methods are available for assessing the degree of risk overall, the most common being the likelihood/severity matrix which uses a three-part scale, high, medium to low.[17] Using available data, the health and safety risks associated with each set of operations has been rated on the following 'high - medium - low' scale:[18]

- high risk [h] - indicates 'significant' health and safety risks above the 1:1,000 threshold
- medium risk [m] - indicates 'tolerable' health and safety risks within the ALARP region
- low risk [l] - indicates 'non-significant' health and safety risks

So in Figure 10, the first column gives the level of risk associated with work operations within organisations generally, based on accident rate data per 100,000 employees. The middle column provides an estimate of the impact of refurbishment activity on these generic levels of operational risk. The last column provides a framework for considering and profiling the range of risks that will need to be managed within a specific project, taking account of all 'case specific' factors that might modify the generic risk estimates, either up or down.

A similar logic is applied in Figure 11, but directed to the generic operations of refurbishment. The first column gives levels of risk

---

17 CIRIA (1995) *Control of risk.* (Report 145) UK: CIRIA and Cox SJ and NS Tait (1991) *Reliability Safety and Risk Management.* UK : Butterworth-Heinemann Ltd.
18 For the organisational profile the risk categories were based on 100,000 per employee rates from the existing HSE tables for 1993/94 all reported injury types: High is <601, Medium is 301-600, Low is >300. For the refurbishment profiles the risk categories were based on reported injury rates from the HSE 1985/86 - 1995/96 process list and converted into 100,000 per employee rates from existing 1985 - 1996 construction employment figures: High >100, Medium is 31-100 and Low is <30. The high rating indicating risk rates exceeding 1:1,000. There are obvious limitations to these data and the risk categories, but in the absence of more definitive data these have to be used at this time.

associated with construction operations, using the HSE data relating injuries to construction employees by work process. [19]

Figure 10   Organisational risk profile

generic operations		available statistical records	generic refurbishment estimate	project specific estimate
input operations	transportation and access	h	h	
	communications	m	m	
	utilities	h	h	
site operations	internal distribution	m	m	
	storage and handling	m	h	
	internal communications	m	m	
core operations	production operations	h	h	
	assembly operations	h	h	
	processing operations	h	h	
	examination and testing	l	l	
	transport services	h	h	
	investigation and research	m	m	
	information services	l	l	
	treatment services	m	h	
	accomodation services	l	h	
	retail service operations	m	h	
	educational services	m	m	
	leisure service operations	m	m	
	other specialist services	h	h	
	facility support services	m	m	
output operations	transportation	h	h	
	communications	m	m	
	distribution of goods	h	h	
	supply of services	m	m	
	disposal of waste	h	h	

[19] The HSE use 43 process categories to classify their RIDDOR data, these have been amalgamated to 21 process categories with the data being reclassified as shown in Figure 11.

Figure 11   Refurbishment risk profile

generic operations		available statistical records	generic refurbishment estimate	project specific estimate
input operations	transportation and access	h	h	
	special services	m	m	
site operations	internal traffic management	m	m	
	delivery and site storage	h	h	
	manual handling operations	h	h	
	mechanical handling	l	l	
	cutting operations	l	l	
	hot work operations	m	h	
	waste storage operations	l	l	
	demolition operations	l	h	
	ground work operations	h	h	
	scaffolding and temp supports	m	h	
	structural erection	m	m	
	services installations	m	m	
	cladding and glazing operations	m	h	
	surface finnish operations	h	h	
	equipment installations	l	l	
	project support services	m	m	
	other processes	h	h	
output operations	transportation	h	h	
	disposal of waste	m	m	

The middle column provides an estimate of generic refurbishment risks. The last column is provided for the adjustment of these generic risks to take account of specific project circumstances. The two risk profiles, derived from the use of Figures 10 and 11, taken together, can be used to evaluate the likely combined impact of all high risk processes, hazards and activities within a specific project, on health and safety overall.

3.6   Risk monitoring and control

Two principal measures are available for the control of hazards, processes and activities that hold significant safety risks. Risks may be 'designed-out' or 'managed-in'. The **designed-out** measures seek to eliminate, replace or modify the hazard, process or activity that would generate significant risk. The **managed-in** measures seek to control the significant residual hazards, operations and activities through strategic, operational and tactical management interventions, and by the adoption of protective measures and contingency arrangements.

Both of these measures assume that the ownership of health and safety risk is clearly understood. In the case of construction projects it rests with the client. Options for risk control from the client's viewpoint are; the elimination of risk, the transfer of risk to others, the reduction of the residual risks, taking or sharing responsibility for all retained risks and the partial avoidance of risk in the sense of not fully complying with the regulatory requirements. The commissioning of a competent professional team by the client, aims to secure an expert approach to the management of risk, and to put in place a combination of appropriate measures, tailored to meet the specific circumstances of the project in hand for the effective control of health and safety risks to acceptable levels. The two main measures for consideration, 'designing-out' and 'managing-in', are described below:

There are three principal ways in which 'designing-out' can assist in risk control:

- **risk elimination**: the elimination of areas of significant risk completely, through the total avoidance of particular hazards, materials, processes and activities.
- **risk substitution**: the reduction of the remaining risks by substituting high risk design concepts, construction methods, site processes and work activities, with others that have lower

levels of risk, both during the refurbishment project and for the remainder of the building's effective life.

- **risk isolation**: limiting areas of significant risk, and their potential impact, by isolating and controlling those that could come into contact with the hazard, process or activity, to reduce the possible consequences associated with the area of risk.

Risk control by 'managing-in' focuses on the residual risks that cannot be 'designed-out' to any significant degree. There are four principal means for reducing these risks:

- **risk transference**: sharing the risk with others through contract terms, reasonable disclaimers, and appropriate insurance arrangements.
- **risk modification**: changing the project risk profile through the introduction of innovatory management systems with better control of site operations, construction processes and procedures, and the day to day working practices that affect health and safety.
- **risk awareness**: gaining familiarity with the site and its buildings, their potential hazards, problems and risks through site induction sessions for all operatives new to the site, and at the start of each new phase of the construction process.
- **risk training**: implementing site and project induction sessions, site operation safety briefings, and safety awareness CPD courses through which a secure construction safety culture can be put in place.[20] These measures are essential when young and less experienced operatives are to be employed.

The active monitoring of health and safety performance throughout a project is an essential component of the risk management process. Under the CDM Regulations there is a requirement that formal monitoring activity takes place. Four main issues must be considered:

---

20 Dedobbeleer and German (1987) found that induction sessions concerning worker safety had a positive impact.

- the statutory requirements for monitoring
- the system that needs to be established to meet the specific conditions and risks of the project in hand
- arrangements for the use of a Health and Safety File to fulfill all CDM requirements
- the adoption of secure methods for auditing health and safety performance overall

Under the RIDDOR legislation, employers are obliged to report accidents in the workplace. The types of accidents and occurrences that are required to be reported to the HSE are outlined and defined in the legislation (HSE RIDDOR, 1996). At present, the HSE estimates that only 50% of accidents in the workplace are reported. In the construction industry the situation is worst. It is estimated that less than 40% of the total construction accidents are formally reported.[21] Improvements in the current understanding of construction safety will rely on the RIDDOR obligations being met diligently and more fully than in the past. Here, it is essential that all reported accidents are investigated, even those involving minor injury, as a part of the monitoring of safety performance. The key objective should be to determine how and why the accidents occur so that safety provisions may be reviewed and management arrangements improved.

The Health and Safety File is maintained to record the possible future health and safety risks that will need to be avoided or contained during the subsequent use of a building and any further modifications and adaptation. The File provides structured information to support the effective management of health and safety risks throughout the remaining lifetime of a building. The use of Health and Safety Files over time should be monitored against accident frequency and safety records. This will be valuable in two ways. First, to understand the effects of the File

---

[21] Waller et al (1989) reported in a USA survellance study that only 31% of accidents were claimed under workman's compensation. Landeweerd et al (1990) indicated in a Dutch study that only 30-40% of lost-time accidents are reported. There is a suggestion that injury claims are likely to raise insurance premiums and are thus not made to avoid the increase.

information on health and safety issues in construction, whether it can be shown to contribute to a reduction in accident frequency and severity or not. Second, to identify those types of accidents that continue to occur, despite the information provided by the Health and Safety Files. Details of the practical duties, responsibilities, measures and procedures for risk monitoring, control and review, are the subject of the next and final part of this *Guide*.

3.7 Safety audit and review

Current refurbishment practice tends to use informal and qualitative methods for risk assessment, conducted on the basis of general experience and judgement alone. While accident databases are maintained, these are rarely referred to during the risk management process. Methods for auditing construction safety and the management of risk are also at an early stage of development. The basis for site safety performance audits has been developed by a team from UMIST for the HSE.[22] Their audit procedures and accompanying checklists, target four main areas of site safety; scaffolding, access to heights, site housekeeping and tidiness, and personal protective equipment. Two important additional areas for the development of audit procedures are the review of near-miss incidents and their causes, and measures for the audit of health and safety management performance itself. The systematic collection of near-miss data would be particularly valuable since it could provide an early warning of recurring problems that might lead to more serious incidents with time. The results of safety audits, conducted at a variety of sites, with reporting to a central body such as the HSE or CIOB, would assist in the improving predictions of high risk circumstances, sites and conditions.

The final activity in the risk management process is the formal review of the relative success of the systems and procedures that

---

[22] Duff et al (1993) *Improving Safety on Construction Sites by Changing Personnel Behaviour.* London : HMSO.

have been used in a project. Health and safety performance should be judged against the Statement of Objectives, as set out at the beginning of the process. Issues to be re-examined and reviewed include:

- the success, or not, of the co-ordination, communication and reporting structures used throughout the project, at strategic, operational and tactical levels
- the sufficiency, or not, of key health and safety performance criteria and standards used within the safety strategies, safety plans and safety actions that have been implemented
- the basis of the company's health and safety Statement of Objectives in the light of experience gained, examining consistency between actions taken and the policies as set
- actions outside of the scope of any individual project, for example, changes in legislation, changes in company policy, changes in user perceptions or social acceptance of accident risks
- the frequency with which future reviews should be carried out

The review should provide feedback to promote corrections and improvements to the risk management system overall.[23] This process by which information of success and failure are fed back into the system, will also require a clear set of procedures concerning the form, timing and level of reporting. The feedback process needs high visibility, with published performance details in an annual report. For any success in the long term, it must be seen to encourage continuity and improvement, project by project, with endorsement at senior executive and board level,[24] ideally integrating health and safety objectives within the general business objectives and plans of the organisation.

---

[23] Sulzer-Azaroff (1987) outlines the successful use of primary feedback loops in industrial safety programmes. Komaki et al (1978) and Nasanen and Saari (1987) provide examples from construction.

[24] Dester and Blockley (1995), HSE (1991), Heinrich et al (1980), Laufer (1987) and Mattila et al (1994) are consistent in reporting that management support is an essential factor in ensuring safety performance.

4 Project co-ordination and communication

4.1 Duties under CDM

All construction projects that are non-domestic, of more than thirty days duration, or that involve more than four workers on site, are 'notifiable' under the CDM Regulations. In addition, all construction projects of whatever scale that are to be carried out in office, retail or similar premises while they remain in normal use, are also subject to notification. Figure 12 shows the appointments that need to be made at the start of a 'notifiable' project. The first crucial step is to establish who is the client, since the responsibility for health and safety rests with them. By the end of 1997 there had been 21 convictions for breaches of the CDM Regulations.[1] In 57% of these convictions, the client was the defendant. The importance therefore, of an unambiguous clarification of who is the 'client' cannot be overstated.

The CDM Regulations define 'client' as "any person for whom a project is carried out, whether it is carried out by another person, or is carried out in-house" [Reg. 2(1)]. So there are two main options; to retain client responsibilities in-house, or to appoint an agent and formally assign client responsibilities to them. An agent is defined as "any person who acts as agent for a client in connection with the carrying out of a trade, business or other undertaking (whether for profit or not)" [Regulation 2(1)]. When responsibility is retained, it rests with the client organisation and

---

1   Summary of convictions provided by the HSE, of the 21 reported convictions to 5 November 1997 - 12 were clients; 2 were planning supervisors; 3 were designers; 3 were principal contractors; and 1 was a contractor.

its facilities management team. When responsibility is assigned, then it lies with the agent who will have sole responsibility overall, for meeting health and safety requirements.

Figure 12 CDM appointments

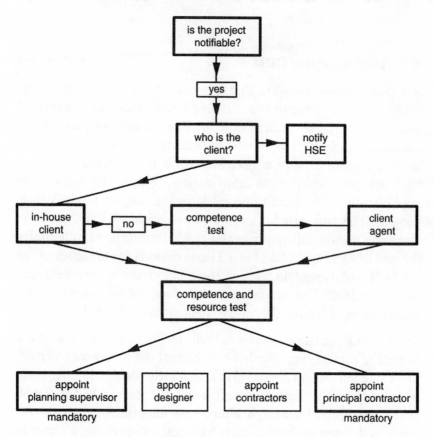

Before appointing an agent, the client needs to consider a number of issues. Employees of the client, who are already undertaking duties on behalf of the client, will not automatically become agents for the client within the remit of the CDM definition. For example, an out-sourced FM provider, can not be considered as the client's agent unless a specific contract defining this role is put in place. Similarly, project managers should not be considered as the client's agent, they will not have statutory responsibility for health and safety matters unless specific undertakings are made.[2] It is important therefore, for the client to be aware that, unless they formally assign their responsibilities to an agent, they will retain liability for breaches under the CDM Regulations.

Once the 'client' for the project has been determined, then they must appoint a **planning supervisor** to co-ordinate and manage the health and safety aspects of the design. The planning supervisor is "any person for the time being appointed under regulation 6(1)(a)" [Reg. 2(1)], with duties detailed in CDM Regulation 14. The appointment of a planning supervisor is a statutory obligation for all notifiable projects. Some 50% of client convictions under CDM relate to their failure to make such an appointment.[3]

As shown in Figure 12, the 'client' may next appoint a **designer** for the project but this is not a mandatory requirement under the Regulations. Designers are "any person who carries on a trade, business or other undertaking with which he: a) prepares a design, or b) arranges for another person under his control (including, where he is an employer, any employee of his) to prepare a design, relating to a structure or part of a structure" [Reg. 2(1)]. The duties of designers are detailed in full in CDM Regulation 13.

The 'client' also has a statutory obligation to appoint a **principal contractor** to co-ordinate and manage health and safety on site. The principal contractor is "any person for the time being appointed under regulation 6(1)(b)" [Reg. 2(1)], with duties as outlined in CDM Regulation 16. The last appointment to be made under the CDM Regulations relates to the **contractor**(s), who is "any person who carries on a trade, business or other undertaking (whether for profit or not) in connection with which he:

---

2   CIRIA Report 172 (1998) discusses the ambiguous role of the project manager under CDM and the need for the client to specify this more precisely. This will ensure that health and safety competence and resource tests would be required for project managers, which at present are not.

3   Summary of convictions provided by HSE.

a) undertakes to carry out or manage construction work, b) arranges for any person at work under his control (including, where he is an employer, any employee of his) to carry out or manage the construction work" [Reg. 2(1)]. The contractor's duties are outlined in CDM Regulation 19.

These five types of appointment, as shown in Figure 12, can involve any party provided that two key requirements are met. First, competency must be demonstrated in relation to the duties that are to be undertaken. Second, the allocation of appropriate resources to support these duties needs to be secured. The Approved Code of Practice requires that reasonable steps are taken in making enquiries as to the competence of intended appointees. The basis for testing competence, in relation to mandatory CDM appointments, is summarised in Table 4.

Competency should be demonstrated in relation to:

- knowledge, experience and understanding of the work involved and of the relevant health and safety standards
- the capacity to apply this knowledge and experience to the work required in relation to the particular project for which the planning supervisor, designer or principal contractor is being engaged

The choice of appointments will result from management decisions on how best to plan, organise, execute, monitor and control health and safety issues during a specific refurbishment project. Here, the facilities manager can often be appointed to act in one of a number of capacities; particularly as 'expert client', or as planning supervisor. This can be especially valuable when occupied buildings are to be refurbished. The facilities manager's knowledge, both of business operations and activities, and of the building fabric, services and conditions, places him in a favourable position for the co-ordination and communication of information throughout the project. The practical duties and responsibilities, as introduced in Chapter 1, are summarised below:

Table 4   Test for competence[4]

CONSIDER as appropriate

Planning Supervisor:
[a]   membership of relevant professional body
[b]   the planning supervisor's knowledge of construction practice, particularly in relation to the nature of the project
[c]   familiarity and knowledge of the design function
[d]   knowledge of health and safety issues (including fire safety issues), particularly in preparing Health and Safety Plan
[e]   ability to work with and co-ordinate the activities of different designers and be a bridge between the design function and the construction work on site
[f]   the number, experience and qualification of people to be employed, both internally and from other sources, to perform the various functions in relation to the project
[g]   the management systems which will be used to monitor the appropriate allocation of people and resources in the way agreed at the time when these matters are being finalised.

Principal Contractors and Contractors:
[a]   the arrangements the contractor has in place to manage health and safety including fire safety
[b]   the procedures the contractor will adopt for developing and implementing the health and safety plan
[c]   the approach to be taken to deal with the high risk areas identified by designers and planning supervisor
[d]   the arrangements the contractor has for monitoring compliance with health and safety legislation
[e]   the people to carry out or manage the work, their skills, knowledge, experience and training
[f]   the time allowed to complete the various stages of the construction work without risks to health and safety; and
[g]   the way people are to be employed to ensure compliance with health and safety law.

---

4   HSC (1995) *Managing construction for health and safety: Construction (Design and Management) Regulations 1994 Approved Code of Practice.* London: HSE.

Client

- establish themselves as client or appoint client's agent [Regs. 4 and 5]
- ensure competency of the proposed planning supervisor, principal contractor, and designer [Reg. 8], and any other appointees
- appoint planning supervisor and principal contractor for the project [Reg. 6(1)]
- ensure adequate resources are available to support the work of the planning supervisor, principal contractor and designer [Reg. 9] and other direct appointees
- ensure that the planning supervisor post remains filled throughout the project [Reg. 6(5)]
- take all reasonable measures to ensure that all relevant information is made available to the planning supervisor [Regs. 11 and 12] concerning:
  - the premises and site
  - the processes and operations being conducted on site
  - the client's Health and Safety Plan [15(3) and Appendix 4 of the ACoP points 3, 7, 8 and 9]
- ensure existing Health and Safety File is kept available for inspection at all times [Reg. 12(1)]

Planning Supervisor

- ensure that notification of the project is given to HSE [Reg. 7(1)]
- ensure designers fulfil their duties under CMD [Reg. 14(a)]
- ensure designers co-operate on health and safety issues [Reg. 14(b)]
- advise client and contractors on competency requirements, and resource availability for designers [Reg. 14(c)]
- ensure that a Health and Safety Plan is prepared for any contractor prior to tendering and includes relevant information [5] [Regs. 15(1), 15(2), and 15(3)]

---

[5] Unless specifically contracted to do so, there is no requirement for the planning supervisor to actually produce the Health and Safety Plan.

- advise client on competency requirements and resource availability for contractors [Reg. 14(c)]
- ensure that a Health and Safety File is prepared and delivered to the client [Regs. 14(d), 14(e), and 14(f)]

Designer

- inform client of CDM responsibilities [Reg. 13(1)]
- ensure that any design considers the health and safety risks involved to workers by avoiding foreseeable risks, combating risks at source, or providing measures for protection against risks [Reg. 13(2)(a)]
- provide relevant design information
- co-operate with planning supervisor and with other designers [Reg. 13(2)(c)]

Principal Contractor

- ensure competence of designers and contractors that they appoint [Regs. 8(2) and 8(3)]
- ensure resources available for designers and contractors that they appoint [Regs. 9(2) and 9(3)]
- ensure that the health and safety plan is maintained during the course of the project [Reg. 15(4)]
- ensure an integrated approach to health and safety management on site [Reg. 16(1)(a)], including the co-ordination of information from contractors [Reg. 19(1)(b]), risk assessments as necessary for construction activities that might affect employees and the public [MHSWR (Reg (3)]
- ensure contractors and employees comply with Health and Safety Plan rules [Reg. 16(1)(b)]
- ensure only authorised persons are allowed access to the construction site [Reg. 16(1)(c)]
- provide planning supervisor with information for the Health and Safety File [Reg. 16(1)(e)]
- ensure contractors are provided with necessary information on health and safety risks [Reg. 17(1)]

- ensure contractors provide employees with information on health and safety risks, emergency procedures and necessary health and safety training [Reg. 17(2)]
- ensure discussion and co-ordination of employee views on health and safety matters [Reg. 18]

Contractors

- ensure competence of designers and contractors that they appoint [Regs. 8(2) and 8(3)]
- ensure resources available for designers and contractors that they appoint [Regs. 9(2) and 9(3)]
- co-operate with principal contractor to comply with statutory duties [Reg. 19(1)(a)]
- provide principal contractor with information for the Health and Safety File [Reg. 19(1)(b)]
- comply with Health and Safety Plan rules [Reg.19(1)(d)]
- provide principal contractor with information required under Reporting of Injuries, Diseases and Dangerous Occurrences Regulations 1995 [Reg. 19(1)(e)]
- provide principal contractor with information for the Health and Safety File [Reg. 19(1)(f)]
- ensure employees and self-employed are provided with the names of planning supervisor, principal contractor and relevant information from the Health and Safety Plan [Reg. 19(2)].

4.2 Project planning

The professional co-ordination of the new duties and responsibilities under the CDM Regulations, coupled with the planning of their application within the specific circumstances of a given project, presents a major opportunity for improvement in health and safety management during refurbishment. In large projects these initiatives may be led by a specialist consultant, under the direction of the planning supervisor, assisted by the project manager. In smaller projects implementation will normally rest with the planning supervisor in matters relating to design, and with the principal contractor for health and safety issues arising from construction operations and activities.

The nature of the project planning task needs to be made clear to all participants. While the measures for project planning and health and safety co-ordination will tend to be unique to each set of project circumstances, the three generic levels of co-ordination - Strategic, Operational and Tactical (see section 2.5), should always be examined as early in the project as is practicable, and before the commencement of work on site. Figure 13 shows these three interface levels for co-ordination and intervention. Planning at the strategic level, will tend to provide the most significant benefits for a project overall. However, strategic level decisions usually have important implications for operational and tactical safety issues. Safety strategies are needed to support safety plans, which in turn help to determine appropriate safety actions.

Figure 13   The three levels of intervention

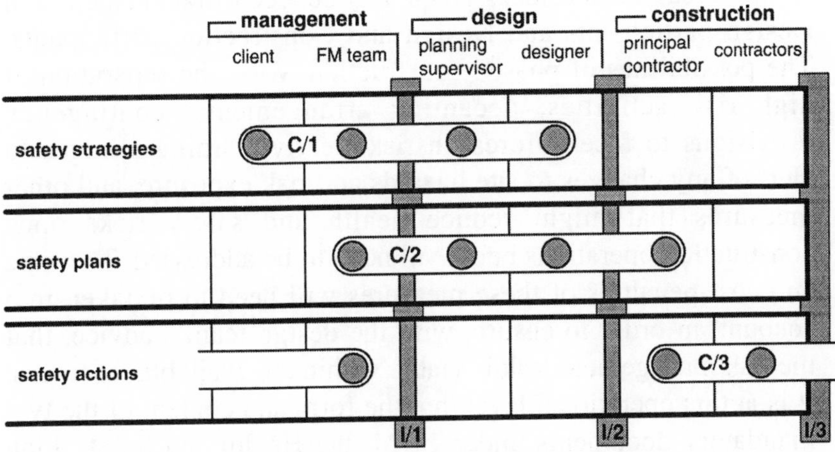

Planning at the strategic level (**C/1**) will involve the client, the FM team and the designer, in the consideration of safety strategies for the project, and for the subsequent use of the building after refurbishment. This ensures that the co-ordination and communication interface (**I/1**) between Management and Design participants is clearly established. From the outset, safety considerations should form an integral part of the design process, with health and safety requirements and strategy clearly stated as part of the strategic brief. When occupied buildings are refurbished the client will require that these strategies do not threaten working continuity. They must not cause undue disruption or put the essential operations of the core business at risk. So the input of the client's knowledge and experience is crucial to interventions at the strategic level. An open review of the client's experience and competence in health and safety matters needs to be undertaken first, and should include any facilities management arrangements that are in place. The outcome of this review will inform decisions concerning the possible appointment of a client's agent and the need or not for specialist advisors.

Planning at the operational level (**C/2**) requires that the client and the FM team, the planning supervisor, designer and principal contractor review project logistics in relation to the anticipated health and safety risks. This secures the co-ordination and communication interfaces (**I/1** & **I/2**) between Management and Design participants and Design and Construction participants. The possibilities of phased construction work, the separation of high risk activities, decanting arrangements, contingency provisions to face unforeseen risks, early warning systems to alert of any changes to site hazards and risk exposure, and other measures that might reduce health and safety risks once construction operations begin, all need to be addressed. The time and cost penalties of these measures will need to be taken into account, in order to ensure, with the design team's advice, that the risk management task is viable within the available resources. It is at this operational level that the form and content of the two mandatory documents under CDM, the Health and Safety Plan and the Health and Safety File, must be considered and agreed. Tables 5 and 6 summarise the essential requirements that will need to be met.

Table 5   Health and Safety Plan[6]

Items for inclusion under Regulation 15 (1) - (3)

[a]   nature of project
-    name of client
-    location
-    nature of construction
-    time scale for completion

[b]   existing environment
-    surrounding land uses and related restrictions
-    existing services
-    existing traffic systems and restrictions
-    existing structures
-    ground conditions

[c]   existing drawings
-    available drawings of structure

[d]   the design
-    significant hazards or work sequences identified
-    structural design precautions or sequence issues
-    reference to contractors' required method statements

[e]   construction materials
-    health hazards arising from construction materials

[f]   site-wide elements
-    positioning of site access and egress points
-    location of temporary site accommodation
-    location of unloading, layout and storage areas
-    traffic/pedestrian routes

[g]   overlap with client's undertaking
-    health and safety issues for project in client's premises

[h]   site rules
-    specific site rules, client or planning supervisor

[i]   continuing liaison
-    procedures for contractors' designs
-    procedures for dealing with unforeseen eventualities

---

6   HSC (1995) *Managing construction for health and safety: Construction (Design and Management) Regulations 1994 Approved Code of Practice.* London: HSE, Appendix 4, see also HSE information sheet no 42, The pre-tender stage Health and Safety Plan and HSE information sheet no 43, The Health and Safety Plan during the construction phase.

Table 6   Health and Safety File[7]

Items for inclusion under regulation 14 (d) - (f). Information contained in the file needs to include that which will assist persons carrying out construction work on the structure at any time after completion of the current project and may include:

[a]   record or 'as built' drawings and plans used and produced throughout the construction process along with the design criteria

[b]   general details of the construction methods and materials used

[c]   details of the structure's equipment and maintenance facilities

[d]   maintenance procedures and requirements for the structure

[e]   manuals produced by specialist contractors and suppliers which outline operating and maintenance procedures and schedules for plant and equipment installed as part of the structure ; and

[f]   details on the location and nature of utilities and services, including emergency and fire-fighting systems

Finally, planning at a tactical level (**C/3**) will involve the client, the principal contractor and the other contractors in assessing the safety implications of all activities, tasks and actions that are planned to take place on site. This establishes the co-ordination and communication interface (**I/3**) between the Construction and Management participants. It must be expected that the client organisation's employees will be unfamiliar with the risks of construction activity. Conversely it is likely that construction workers will be unfamiliar with the operations and activities of the building occupants. They may be unaware of the risks to health, safety and human wellbeing that their construction activities might imply. Planning will seek to eliminate problems and risks wherever possible, particularly those where occupant

---

[7]   HSC (1995) *Managing construction for health and safety: Construction (Design and Management) Regulations 1994 Approved Code of Practice.* London: HSE, Appendix 5 and HSE information sheet no 44, The Health and Safety File. Also for building services, Nanayakkara, R (1997a) *Standard Specification for the CDM Regulations Health and Safety File.* (BSRIA AG 9/97) UK: BSRIA.

activities and construction activities might coincide in space and time. Where this is not possible, overlaps should be reduced in extent and reduced in time, to minimise risk. The planning of overlapping activities should be fully completed prior to tender, and formally recorded in the contract documentation.

The preferred option in all planning initiatives should always be the elimination of risks from the outset, by 'designing-out', 'planning-out' and 'organising-out' the areas of significant risk (see section 3.6). Where this is not possible, then residual risks must be 'managed-in' through the control and monitoring phases of the risk management process. The use of the CDM decision framework across each of its three interfaces will help to ensure that the issues to be addressed will be understood and shared by all participants. The risks and duties at each of the three intervention levels will next be detailed in turn.

4.3 Safety strategies

The first Interface (**I/1**) provides the means for project co-ordination at the strategic level, and deals with the relationship between **management** (client, facilities manager, client's agent) and **design** (designers, planning supervisor). This interface links the business operations and activities of the organisation, the health and safety of its employees, customers and suppliers and others that may be put at risk by the refurbishment work, to the provisions of the design proposal, its specification and implementation.

The Design - Management Interface is shown in Figure 14, where the horizontal bar represents the scope of co-ordination and the dots indicate the principal participants that are involved. Naturally, the scale of the project will affect the degree to which any particular type of participant will become involved. At this level of co-ordination the client looks for the strategic integration of health and safety concerns by the design team and the planning supervisor. If possible, the appointment of the principal contractor should also be made at this stage, particularly if the project is complex, or is planned to be undertaken in phases. The early involvement of the principal contractor will help to address the issues of buildability which can have a significant impact on strategic level decisions for health and safety.

The designers must examine their design proposals and all of the health and safety issues that they imply. They need to ensure that the design concept, its form, structure and materials, its environmental regime and services, its buildability, the intended stages and methods of construction, and the building's subsequent use, cleaning and maintenance, do not give rise to unnecessary risks to human health and safety.

Figure 14   The Design - Management Interface

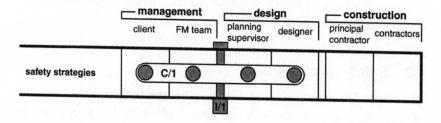

The CDM duties for the key participants at the strategic level arise during the early stages of a project, typically during the feasibility, briefing and design stages of work. In refurbishment projects they normally focus on the unimpeded and safe continuation of the organisation's operations, and on the safety of construction processes on site, once refurbishment begins.

For the **client** the practical actions that need to be considered at this stage are:[8]

• the appointment of an agent to represent them for the project depending on the competence, or not, of the available internal

---

[8]   see also CIRIA (1997) *CDM Regulations - Practical guidance for clients and clients' agents*. Report 172 London: CIRIA and HSE information note 39, The role of the client.

personnel. In refurbishment work, it is important that an agent has powers to act and is available on site so that decisions can be taken and implemented immediately as required. These responsibilities must be clearly defined in the agent's contract of appointment

- the test of competency for the planning supervisor, designer and principal contractor as described in section 4.1 and summarised in Table 4. Key issues to consider are:

  - the lack of specific safety data to inform risk assessments, places reliance on the use of professional judgement and subjective opinion. Experience with similar refurbishment projects for all of the participants is therefore desirable

  - the planning supervisor's understanding of the design process, his or her practical on-site knowledge of the construction process, and of construction health and safety

  - active management involvement in accident prevention measures, reporting and control procedures, and accident review and monitoring arrangements by the principal contractor. Management commitment is one of the prime determinants of the success of health and safety practices on site

- the availability of resources, both time, finance and personnel, devoted to the project in hand. Time pressures are a contributory cause of accidents that are outside the control of individual contractors yet have a direct impact on injuries on site [9]

- full information about the physical characteristics of the existing building should be compiled prior to work on site and may involve full or selective surveys of the areas to be refurbished. The location of any harmful substances (eg. asbestos[10]) are best considered during this early stage rather than requiring the principal contractor to investigate during their contract period

---

9   Salminen (1994) and Hubbard and Neil (1986) found that time pressures increased the accident rates on site.

10   29% of convictions to November 1997 were for client's and planning supervisor's failure to provide information about presence of asbestos.

- the confirmation of the Health and Safety Plan details by the client to ensure that all parties are aware of what is expected, specifically:
  - the availability of existing drawings (and Health and Safety Files)
  - any overlap of client and contractor undertakings and how these are to be isolated
  - site rules for dealing with client's employees, public access and site traffic management
  - clear and continuing liaison procedures related to any design changes that the client or other members of the professional team might make.
- the confirmation of the Health and Safety File details by the client to ensure that all parties are aware of what is expected.

For the **planning supervisor** the main procedural issues for consideration at this stage are:[11]

- ensure notification of the project to HSE
- ensure that designers are fully aware of, and fulfil their duties. Expectations should be clearly set out at the beginning of the project as to what will be required from all parties, with agreed arrangements for site visits and periodic project reviews
- the previously mentioned competency issues should be examined in detail

For the **designers** the main actions at this stage are:[12]

---

[11] see also HSE information sheet no 40, The role of the planning supervisor, and CIRIA (1998) *CDM Regulations - Practical Guidance for Planning Supervisors*, Report 173. London: CIRIA.

[12] see also HSE information sheet no 41, The role of the designer; Bone S (1995) *Information on site safety for designers of smaller building projects*. UK: HMSO; CONIAC (1995) *Designing for health and safety in construction: A guide for designers on the Construction (Design and Management) Regulations 1994*. London: HMSO; CIRIA (1997) *CDM Regulations - Work sector guidance for designers*. (166) London: CIRIA; and Haverstock, H (1996) *The building design easibrief CDM primer*. UK: Miller Freeman plc.

- ensure that the client has a full understanding of their responsibilities, especially for providing all available information about the existing building
- demonstration how health and safety risks are to be considered and comparisons made between alternative courses of action. This typically involves undertaking a risk assessment, beginning with a preliminary review of the risks listed at the end of this section. These risks need to be considered throughout the design stage with emphasis reflecting the specific circumstances of the building and site.

Should the **principal contractor** be appointed at this early stage of the project, he should:

- advise on competence of designers and contractors
- advise on the issues of buildability and hazardous operations, for example, demolition or asbestos removal
- review the measures being taken for an integrated approach to health and safety from a constructional and working logistics viewpoint, indicating how this could affect the allocation of resources

The consideration of safety strategies to combat the identified health and safety risks at the Design - Management Interface, will have different priorities for the various participants in the process. For the **client**, in addition to any specific health and safety risks that may be encountered, the practical success in containing project risks will be affected, to a considerable extent, by the degree to which the intent and commitment of the client is visible to all. The more positive the approach, the better health and safety on site is likely to be. The nature of both the construction industry and the requirement for competitive tendering, suggests that safety improvements are most likely to occur in those projects were the client is seen to be leading health and safety initiatives (see Case 1).[13] This commitment begins with the Health and Safety Policy and continues through to the allocation of appropriate resources to project safety.

---

13 Simonds and Shafari-Shafari (1977) reported that lower injury rates are related to visible top management involvement in health and safety.

Case 1

An example of the client taking the lead for site safety within a construction project: the Glaxo Group Research Headquarters Building Project had visible client commitment and support for site safety. The client determined the safety agenda and made sure that appropriate resources were available to implement the plans. As a result the project was finished within the tight timescale, it was recognised as the safest construction site in Britain that year.[14]

The key areas for strategic review by the client concern:

- the **business plan** and working procedures to identify any areas of potential conflict between the organisation's business objectives and the execution of the refurbishment project
- **core operations** and facilities management procedures to see where key overlaps are likely to occur between occupants' activities and refurbishment operations (see Case 2)

Case 2

In a hospital refurbishment project the phasing of activities was reviewed at a strategic level through a User Panel. The Panel included representatives from the major client departments, facilities specialists and the design team. Given the lack of specific data on all of the risks that the refurbishment project might involve, additional methods for determining risks were required. The meetings of the User Panel provided a qualitative assessment of the identified risks and a collective and clearer understanding of the implications of the work was developed. [15]

---

[14] Winch et al (1997) Towards total project quality: A gap analysis approach *Construction Management and Economics*. vol. 18 no. 2.

[15] Bickerdike Allen Partners, unpublished case study prepared for "Health and safety risk analysis for refurbishment of occupied buildings." CMR LINK project 323.

- **time and resource** pressures and constraints
- **building information** issues such as:
  - existing fire safety systems and egress points, and the measures for continuing services while works are in progress
  - existing structural systems within the building
  - the internal and external location and specification of all major services, the existing external electrical service location being particularly important when groundworks are contemplated. Building services for which information must be made available include:
    - mechanical services
    - electrical services
    - water services
    - telecommunications services
    - data services
    - life safety services systems
    - other services
- **health and safety procedures** including fire safety, egress routes, hot works permits systems and their enforcement [16]

For the **planning supervisor,** the **designers** and the **principal contractor** the identified health and safety project risks to be considered at the strategic stage include:

- **phasing** and **decanting operations** for reduced health and safety risk but without undue disruption to the employers' operations
- **planning site activities,** particularly the control, movement and transfer of goods on the site[17]
- **construction works** and the implications for:
  - demolition works proposed

---

16 Croner's (1997) *Management of construction safety* UK: Croner's Publication Ltd, outlines many of these best practices. In addition, the HSE publish guidance notes for all of these areas.

17 Niskanen and Lauttalammi (1989) suggest this is a key to productivity growth in the construction industry. HSE (1992) *Manual handling. Manual Handling Operations Regulations 1992: guidance on the regulations.* (HSE L23) London: HMSO.

- major service interruptions from works
- existing emergency procedures and means of egress
- temporary works requirements, fire exits, etc.

- **demolition** of parts of the existing structure, with particular attention to isolating dust, noise and dirt and to the disposal of the waste material [18]
- **other risks** to health and safety also include buildability and contractual constraints to the time and costs of the project
- **generic process risks** that have been identified for refurbishment projects need to be reviewed in terms of the possible impact of the design on construction safety, for:
  - **surface finish operations** involving carpenters, plumbers, electricians and labourers
  - **manual handling operations** particularly transferring goods on site, involving labourers, managerial staff, carpenters and other occupations on site [19]
  - **groundwork operations** [20] and **surfacing roads** involving paviours, labourers, management staff and groundworkers
  - **other non-construction** processes taking place on site involving labourers, other non-construction staff, and manual workers
  - **structural erection** processes, particularly roofing, involving roofers, labourers and carpenters
  - **delivery** and **site storage** processes, particularly loading and unloading of materials involving labourers, drivers and paviours

---

[18] HSE (1988) *Health and safety in demolition work.* (Guidance Note GS 29) UK: HSE and BSI (1982) *Code of practice for demolition.* (BS 6187) UK: BSI.

[19] Stubbs and Nicholson (1979) analysed 148 manual handling accidents in the construction industry using HSE data from 1971-1976. Analysis indicated that 16-30 year-olds suffer the greatest number of accidents from manual handling (13.0/1000), 50.3% of manual handling accidents were caused by lifting / loading and lifting / carrying.

[20] BSI (1981) *Code of practice for earthworks.* (BS 6031) UK: BSI. Hinze and Gambatese (1996) reviewed US OSHA database from 1985-1989 which suggests trenching risks are greatest at lunchtime and for trenches less than 5 foot wide by 5 foot - 10 foot deep.

Finally, a simple summary of the key considerations in the development of Safety Strategies at the Design - Management Interface is shown in Table 7. These issues give emphasis to the importance of the project planning activity, and to the key role of the client in setting the objectives for the project from a health and safety perspective. At the very least the co-ordination of Safety Strategies should enable an appropriate clarification and definition for the next level of decision, the co-ordination of operations and the development of Safety Plans, interface (**I/4**). It must also anticipate the safety issues that will be encountered during the Safety Action phase.

Table 7 Safety strategies

Consider:

- the business plan of the organisation
- operational health and safety policy and practice
- FM arrangements and procedures
- the stategic brief for the project
- appointments, roles responsibilities of:
  - design team
  - planning supervisor
  - lead consultant
  - specialist health and safety consultant
  - project manager
  - principal contractor

All must enable appropriate definition of:

- the Safety Plan, and
- the Design - Construction Interface

4.4 Safety plans

The second Interface (**I/2**) deals with the relationship between **design** (designers, planning supervisors) and **construction** (principal contractor, contractors). In addition, the client or their agent, will need to be involved in the co-ordination of safety issues during the design, specification and implementation phases of the project. The Design - Construction Interface is shown in Figure 15, it focuses on the potential impact of operational arrangements and site logistics on health and safety issues. The risks associated with each stage of work must be examined, specific high risk activities identified, and all operations and tasks co-ordinated, one with another, for the containment of health and safety risks within the project overall.

Figure 15   The design - construction interface

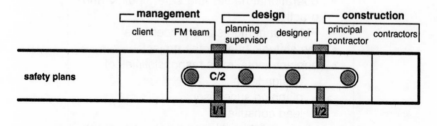

This stage relates specifically to the issues of planning and developing the Health and Safety Plan, as the primary co-ordinating document (see Table 5). It involves the co-ordination of all operations, the pre-tendering activities for health and safety and the documentation that needs to be completed prior to the commencement of work on site. The plan should fully anticipate the tactical issues that may be encountered when work on site begins, and the safety actions that will need to be considered at this level of decision. A summary of the key considerations in the development of Safety Plans at the Design - Construction Interface, is given in Table 8.

Table 8   Safety plans

> Consider:
>
> - the Health and Safety Plan
> - contractual arrangements
> - sectional completions
> - prelims
> - sub-contractor arrangements
> - contractor's design portion supplement
> - specialist contractors
> - performance specified elements
> - temporary works
>
> The Safety Plan must enable:
>
> - the Safety Actions, and
> - the Construction - Management Interface

The duties to be undertaken during the operational stage are, in many cases, a reiteration of those at strategic level but with emphasis on practical implementation rather than option appraisal. For the **client**, the main duties are:

- test of competency for principal contractor and contractor, considering management involvement; specifically the procedures for preventing accidents, reporting and control procedures for reviewing and monitoring accidents should they nonetheless occur, both for the principal contractor and contractors
- ensuring resource availability for designers and principal contractors
- determining how the planning supervisors role will remain active throughout the course of the project

For the **planning supervisor** the main procedural issues to be considered are:

- development of the Health and Safety Plan as agreed with the client

- development of the Health and Safety File as agreed with the client, depending on the terms of the contract. In many cases the planning supervisors role will be passed on to the principal contractor, who will consolidate developments to the Health and Safety Plan during construction and facilitate the co-ordination of information for the Health and Safety File

For the **designers** the main procedural issues to considered at this stage concern the operationalisation of the design, with further clarification of the health and safety risks covered in the previous section, demonstrating how they are to be implemented during the construction process.

For the **principal contractor** the required duties include:

- ensuring that the Health and Safety Plan is developed, maintained and updated during the course of the project and that this information is made available to all relevant contractors. This is one of the main reasons for the dual role of principal contractor and planning supervisor during the construction phase.
- conducting risk assessments, in addition to those conducted by the designers, for the scope of work identified and particularly for construction activities that might adversely affect employees and the public. These should be based on the identified risks from Chapter 3 and are detailed at the end of this section
- ensuring an integrated approach to health and safety management on site, including the co-ordination of information from contractors, induction briefings and training sessions. These measures are especially useful when building and sites are complex, or the project involves multi-phased operations
- ensuring that contractors and employees comply with the rules of the Health and Safety Plan
- ensuring that only authorised persons are allowed into construction site
- providing planning supervisor with information for the Health and Safety File. This process is simplified when the principal

contractor takes over responsibility from the planning supervisor during the construction phase

- ensuring that contractors are provided with necessary information on health and safety risks. This typically involves a review of the relevant sections from the Health and Safety Plan and the principal contractor's assessment of any additional risks
- ensuring that contractors provide employees with information on health and safety risks, emergency procedures and necessary health and safety training
- ensuring discussion, review and co-ordination of employee experience and feedback on health and safety matters

For the **contractor** the duties that are required include:

- co-operating with principal contractor to comply with all statutory duties
- providing principal contractor with information for the Health and Safety File
- complying with the rules of the Health and Safety Plan
- providing principal contractor with information required under Reporting of Injuries, Diseases and Dangerous Occurrences Regulations 1995 [21]
- providing principal contractor with information for the Health and Safety File
- ensuring employees and self-employed are provided with names of planning supervisor, principal contractor and relevant information from Health and Safety Plan

In refurbishment the continued involvement of the **FM team** throughout the operational stages can be of considerable benefit, since they are aware of the existing conditions and are likely to have a view based on experience on how particular design solutions may work in practice. For the FM team, this on-going involvement is also important to review the likely impact of the phasing plans on the support service operations such as catering and cleaning. In addition, a review of the long-term health and

---

[21] HSE (1996) *A guide to the Reporting of Injuries, Diseases and Dangerous Occurrences Regulations 1995.* (HSE L73) London: HMSO.

safety implications in relation to maintenance and cleaning can be an important part of the role that the FM team is able to play.

For the **planning supervisor** and the **designers** the health and safety risks at the strategic level should be reviewed again from an operational perspective to see if residual risks might still be designed-out altogether, or the extent to which control of these risks could be managed-in during the construction process, and as the building is used, maintained and cleaned (see Case 3).

Case 3

> The hospital project team designed a window system to facilitate both the ventilation requirements and the cleaning requirements of the client. The tilt and turn frame enabled the client's window contractor to clean the external glazing and frames from inside the building. [22]

For the **principal contractor** and **contractors** the identified risks that need to be considered are similar to those reviewed in the previous section. The main issue to be considered is the separation of employer operations from construction operations in terms of time, space and 'conditions', as summarised below:

- **time separation**: employer operations and contract operations should ideally not take place concurrently, but complete time separation is usually impracticable. Clear operational procedures must be put in place in all areas where employer and contractor activities might coincide in time, particularly when public access is involved. There are two aspects to this:
  - **work activities** - the first level to consider is the areas of overlap with the employer's operations, particularly if they involve the public or the works are external. Working out-of-operating hours, while an additional expense, can help to ensure that construction work activities do not disrupt the employer's operations unduly, or increase the risk to the public. The second level for consideration concerns the

---

[22] Bickerdike Allen Partners, op. cit.

overlap between different skills and trades within the construction operations. Trade overlaps present additional risks during refurbishment projects, where this is often an unavoidable and common occurrence

- **traffic flow** - the review of the timing of vehicular and pedestrian movement through or around the site is especially important, particularly during normal delivery periods, the peak times of arrival and departure from existing car parking areas and access points to the external road network. Temporary traffic calming measures, such as the use of speed bumps or traffic lights, can be necessary in some cases (see Case 4).

Case 4

> The hospital refurbishment required the use of temporary traffic calming bumps to ensure that the staff did not exceed the speed limit when using the internal roadways and car park. This was done after site work had begun, following reports of several near-misses involving staff and contractors. This demonstrates the importance of monitoring conditions and acting on the findings of safety monitoring arrangements. [23]

- **space separation** is especially important in relation to:
  - **work activities** space being made available for the storage and movement of goods around the site
  - **traffic activities**, where a review of the location of the vehicular and pedestrian circulation routes should be undertaken to ensure the least amount of overlap in circulation pathways as possible
- **conditions separation** involves ensuring that dust, noise and vibration are contained within the construction operations area from which they originate. Since employees do not typically use personal protective equipment the employer operations need to be secured against infiltration. This may require

---

[23] Bickerdike Allen Partners, op. cit.

higher than normal rates of dust extraction to protect third parties. There are several guidance sources for these issues. In addition to obvious health risk from asbestos, dust from cement, stone and wood should also be contained in the construction area of origin (see Case 5).

Case 5

> The refurbishment of Brent Cross Shopping Mall was the first of its kind in the UK. The separation of the site activities raised an interesting dilemma as managing agents and contractors discovered that the public did not want to see demolition work taking place as they perceived such activity to be risky. They did however want to see progress of the refurbishment work. The work was phased to reveal the completed sections of the interior to the public as work progressed.

The management of site wide issues to be considered include:

- **site safety induction** – the principal contractor should organise the information for on-site induction safety with all contractors and relevant employer operations (see Case 6).

Case 6

> Special induction sessions were undertaken for all of those involved in the hospital project works. These were organised following the appointment of the principal contractor to brief staff. This specially targeted precaution was thought to be necessary because of hazards arising from work in hygiene-sensitive areas and for hospital staff to appreciate and understand the possible impact and implications of the contractors remedial works for their normal activities. [24]

- **scaffolding** – Consideration of the use and access to scaffolding works, with measures for the protection of the

---

[24] Bickerdike Allen Partners, op. cit.

public when scaffolding is near or over public footpaths or roadways, with strict control over both the hours of work and the hours of access when this is the case (see Case 7) [25]

Case 7

> The hospital project reviewed the access to and from the scaffolding as part of the consideration for the size of the rainscreen cladding panels. This performance review ensured that the size and weight of the panels were such that reasonable safety was secured in their handling and movement in and around the scaffolding. [26]

- **access to heights** – falls continue to be the basic cause of the majority of fatalities, serious accidents and major injury, irrespective of the construction activity taking place. The available guidance in this area should be reviewed, and is especially critical when employer operations are undertaken adjacent to the construction operations, or when openings are created in existing structures, for example, during lift installations or repairs to stairways
- **housekeeping** – the simple statement that "a tidy site is a safe site," continues to be shown to be the case.[27] Consideration of the delivery and disposal of waste materials needs to be considered where the employer operations have particular requirements or statutory obligations, hospitals are a particularly important example (see Case 8)

---

[25] BSI (1993) *Code of practice for access and working scaffolds, and special scaffold structures in steel*. (BS 5973) UK: BSI. and BSI (1990) *Code of practice for temporarily installed suspended scaffolds and access equipment.* (BS 5974) UK: BSI and as well as HSE publications.

[26] Bickerdike Allen Partners, op.cit.

[27] Nasanen and Saari (1987), Mattila et al (1994) Jannadi (1996), and Duff et al (1993) all suggest that good housekeeping promotes good site safety.

Case 8

> The hospital refurbishment required a clear understanding of the disposal system. The disposal of sharp objects is of particular concern in this environment and a clear understanding of the waste disposal system used is essential to avoid injuries that could seriously damage a worker's health unknowingly. [28]

- **personal protective equipment** – the use of PPE for reducing both the severity of any incident and minor injuries is well understood [29]

The construction processes that have particularly high accident frequencies, as highlighted in Chapter 3 and detailed in Appendix A2, should be examined in relation to the case specific circumstances and conditions of a given project. These generic high-risk processes in refurbishment should be reviewed in terms of the impact of design on construction safety in relation to:

- **surface finish operations** involving carpenters, plumbers, electricians and labourers
- **manual handling operations** particularly transferring goods on site involving labourers, managerial staff, carpenters and other occupations on site
- **groundwork operations** and **surfacing roads** involving paviours, labourers, management staff and groundworkers
- **other non-construction** processes taking place on site involving labourers, other non-construction staff and manual workers
- **structural erection** processes, particularly roofing, involving roofers, labourers and carpenters
- **delivery** and **site storage** processes, particularly loading and unloading of materials involving labourers, drivers and paviours

---

[28] Bickerdike Allen Partners, op. cit.
[29] HSE (1992) *Personal protective equipment at work Regulations 1992: guidance on the regulations.* (HSE L25) London: HMSO.

Finally, the following generic site hazards may present serious safety and health risks, and will need to be reviewed within the context of the particular project in hand:

- **dust** created from demolition or repair work should be contained within the work area with checks to ensure that it is not dangerous to either the construction staff or the staff in occupation. Dust control is a particular problem for those dealing with insulation, wood and silicious materials (see Table A27).[30] Asbestos based materials constitute a major and significant health hazard.[31] All involved with work associated with this material must consult the appropriate codes of practice

- **noise and vibration** which, depending on its extent and persistence, can interrupt the work of staff, disrupt the client's business, and cause nuisance to members of the public, as well as increasing health and safety risk. Here the best remedy is the reduction of the nuisance at source (see Case 9) [32]

Case 9

The hospital project contractors used a water jet for concrete cutting, as it was found that this method was the least intrusive in relation to be noise and vibration, when remedial repair works are undertaken.[33]

---

[30] Ferguson I (1995) *Dust and noise control in the construction process.* (HSE CRR 73/1995) London: HMSO and Health and Safety Executive (1995) *General COSHH (Control of substances hazardous to health).* (HSE L5) London: HMSO.

[31] HSE (1993) *The control of Asbestos at work. Control of asbestos at work Regulations 1987: Approved Code of Practice.* (HSE L27) London: HMSO and HSE (1993) *Work with asbestos insulation, asbestos coating and asbestos insulating board. Control of asbestos at work Regulations 1987: Approved Code of Practice.* (HSE L28) London: HMSO.

[32] BSI (1984) *Noise control on construction and open sites: Part 2. Guide to noise control legislation for construction and demolition, including road construction and maintenance* (BS 5228:part 2) UK: BSI.

[33] Bickerdike Allen Partners, op. cit.

- **pollution and rubbish**, which are detrimental to staff, other occupants and the public, and can constitute a serious health hazard in extreme cases, need to be controlled through work planning and good site housekeeping
- **fire** represents a serious potential hazard in refurbishment work, particularly when 'hot works' activity is undertaken in roofs and interiors. All activities with high fire risks should be carefully controlled through a permit system by the FM team, with related method statements required for these activities from all contractors. The BEC Guide: *Fire prevention on construction sites*, is the industry standard to consult on this matter
- **pest infestation** can occur when existing habitats are disturbed by construction activity, typically with rodents, creating health problems
- **unforeseen hazards** need to be anticipated during refurbishment works and tackled through contingency planning arrangements for handling unexpected hazards, should they arise as the existing building fabric is opened up

4.5 Safety actions

The third Interface (**I/3**) relates to the detailed working processes, activities, tasks and safety actions on site. This interface deals with the relationships between the **construction** process (principal contractor, contractors) and **management** (client, FM team, client's agent). The Construction - Management Interface is shown in Figure 16. It is the focus for liaison between the client and the contractors, with emphasis on the safety of construction activities and actions as they are undertaken, day by day, and hour by hour. It is important that the liaison is two way and is clearly defined between those directing the construction processes and those managing the occupant's operations, so that both can run in parallel without imposing undue risk, one on the other.

Figure 16   The Construction : Management Interface

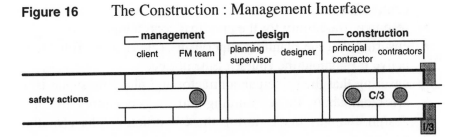

Safety issues at the tactical level rely on measures that can remedy short-term problems as they arise, lessening their impact on the health and safety of individuals, controlling the risks associated with particular sequences of activity, and initiating flexible response to the variable local conditions under which work has to be conducted. Many of the potential problems can be anticipated and planned at an early stage, typically those relating to the isolation of high risk construction activities, the avoidance of organisational and construction operations that overlap in time and space, and contingency measures for facing different weather conditions, unforeseen events and chance occurrences. The principal contractor and contractors need to make their own assessment as conditions and circumstances change, adjusting their working practices accordingly, within the framework of the Health and Safety Plan. A summary of the main considerations in establishing a framework for Safety Actions at the Construction - Management Interface is given in Table 9.

The active involvement of management is one of the prime determinants of the success of practical health and safety measures on site. For the **client** the main CDM issues to consider are:

- test of competency for principal contractor and contractor is outlined in HSC's Approved Code of Practice (see Table 2). The key issues to be considered are management involvement, specifically in clarifying procedures for accident prevention, agreement on reporting and control procedures for reviewing

and monitoring accidents by the principal contractor and contractors
- resource availability for the principal contractors
- ensuring that the planning supervisors role remains filled and active throughout the course of the project
- ensuring that the principal contractor and planning supervisor are clear about the continuing liaison procedures for any design changes made during the construction phase of the project

Table 9   Safety actions

Consider:
- construction operations and processes
- employer operations and processes
- physical conditions and restrictions
- precautionary measures
- potential overlaps and adjacencies in terms of:
  - space
  - time
  - conditions
- Health and Safety Plan revisions
- contracts
- liaison between employer and contractor
- Health and Safety File

For the **planning supervisor** the key issue is practical clarification of what he or she has been contracted to do. As yet there is no 'common' contractual practice as to how the planning supervisor's role should be conducted in detail, and some have suggested that duties might be assigned formally to the principal contractor during the construction phase. The CDM duties are to:

- advise the client and contractors on the competence of designers to carry out refurbishment works
- advise client on the competence of contractors to carry out refurbishment works

- co-ordination of the Health and Safety File information as agreed with the client

For the **designers** the degree of involvement during the construction stages will depend upon the quality of the original design, its completeness when work begins on site and the competency of design communication within the project as a whole. If the design is complete, then the designers formally have no further duties under CDM. If not, then the duties previously mentioned still apply. This eventuality should be considered during the earlier strategic stage at Interface (**I/1**) so that the roles and relationships to face the 'incomplete design' situation are clearly understood by all from the start.

For the **principal contractor** the duties that are required include:

- ensuring competence of designers and contractors
- ensuring resources available for designers and contractors
- ensuring that the health and safety plan is maintained during the course of the project and that this information is made available to all relevant contractors
- conducting risk assessments for the scope of work identified, in addition to those conducted by the designers
- ensuring an integrated approach to health and safety management on site, including the co-ordination of information from contractors and, where necessary, additional risk assessments for construction activities that might adversely affect employees and the public
- ensuring contractors and employees comply with Health and Safety Plan rules
- ensuring only authorised persons are allowed access to the construction site
- providing planning supervisor with information for the Health and Safety File
- ensuring that individual contractors are fully briefed and provided with all necessary information on health and safety risks
- ensuring contractors give their employees necessary health and safety training, and provide them with information on health and safety risks and emergency procedures

- ensuring discussion, co-ordination and feedback of employee views on health and safety matters

For the **contractors** the duties that are required include:

- co-operating with principal contractor to comply with statutory duties
- ensuring competence of designers and contractors, and that appropriate resources are available
- providing principal contractor with information for the Health and Safety File
- complying with Health and Safety Plan rules
- provide principal contractor with information required under Reporting of Injuries, Diseases and Dangerous Occurrences Regulations 1995 (RIDDOR)
- ensuring that employees and any self-employed personnel are provided with names of planning supervisor, principal contractor, together with relevant information from Health and Safety Plan

For the **FM team** most of the risks will have been considered by this stage of the project. The remaining duties normally involve liaison with the principal contractor to ensure that any unforeseen problems are dealt with quickly and effectively. When complex buildings are refurbished, the FM team can be an invaluable source of specialist advice on a day to day basis as work proceeds (see Case 10). The additional safety risks that may arise at this stage of the project are:

- **time** constraints imposed on the project should it get behind schedule. health and safety risks are generally increased when construction activities are conducted under extreme time pressures
- **unexpected hazards** encountered by contractors new to the site. All new and temporary workers to the site should undergo induction or training session that clearly outlines the key hazards of the site, for example traffic flows, scaffolding work areas, openings in slabs, cladding and roofs. This is especially important for those workers who do not have a

particular occupational trade and whose work may occur in many areas of the site

Case 10

The hospital project had regular liaison meetings with the facility management staff. The FM team was an important source of information and expertise on the building and its specialist systems. The facilities management's specialist contractors were used to make the necessary alterations to the hospital's services. The FM team were used as sub-contractors for temporary works preparations for the phased decanting and department relocation.[34]

4.6 Managing health and safety risks

The introduction of the Construction (Design and Management) Regulations 1994 has put in place a secure framework to support the management of health and safety risks in construction. The Regulations provide the opportunity for continuity and consistency in the management of human health and safety, both during the construction of new buildings and throughout the life of the building stock as maintenance, modification, refurbishment and adaptation occurs. The CDM Regulations are however at an introductory stage of application, and the construction industry is still learning how to make use of their provisions effectively. At this early stage, the benefits of CDM are becoming acknowledged more widely but it is to be expected that some drawbacks and difficulties remain and need to be overcome.

The research on which this *Guide* is based, included a detailed review of published health and safety material and documented 'best practice' experience, backed up by a comprehensive examination of all available data concerning the factors that have contributed to industrial accidents in general and construction accidents in particular, since the 1972 Robins Report. This

---

[34] Bickerdike Allen Partners, op. cit.

review was supplemented by structured interviews of health and safety experts and advanced practitioners.[35] Their responses suggested that many were still not fully aware of the implications of CDM and the benefits that it can bring. A considerable degree of consensus was suggested by the review, between those with legal, medical, public health and construction industry perspectives, concerning the benefits and shortcomings of CDM.

Perceived **benefits** included:

- heightened awareness of the health and safety risks in construction throughout the full life cycle of buildings
- clearer interfacing arrangements, liaison roles and lines of communication and co-ordination, for all involved in the production, maintenance and renewal of building stock
- the introduction of 'expert' risk assessment procedures for the evaluation of health and safety risks, particularly in larger projects
- recognition of the designers' contribution to human health and safety in architecture, through risk assessment procedures brought forward into the briefing and design process
- improved quality of relevant health and safety information provided by the client
- a growing tendency for contractor involvement and input to design decisions
- explicit tendering documentation for contractors

Perceived **shortcomings** included:

- emphasis on 'paperwork safety' rather than 'people safety', with excessive reliance on extensive documentation and unnecessary detail, concerning health and safety matters
- the role of the planning supervisor, the requirements for the Health and Safety Plan and the Health and Safety File are still not fully and clearly understood by some

---

[35] These views are also supported by the subsequent HSE review: The Consultancy Company (1997) *Evaluation of CDM Regulations-Final Report* (HSE CRR 158) London: HSE and in CIRIA (1997) *Experiences of CDM*, (CIRIA Report 171) London: CIRIA, and CIRIA (1998) *Implementing CDM*. (CPN Workshop Report 803L) London:CIRIA.

- the presumption that the client will recognise the benefits of project planning and co-ordination and will accept the additional time and cost of CDM compliance
- the lack of systematic consideration of 'real' accident data and the collection of 'near-miss' information to inform hazard identification and health and safety risk assessment
- risk assessment methods tend to be subjective and informal
- the costs of implementing the legislation for both smaller organisations and for smaller projects is seen to be excessive

The challenge now is to build upon the proven benefits and successes of CDM, learning how to overcome the perceived shortcomings, so that the opportunities offered by CDM arrangements can be fully exploited by all. Table 10 provides a concluding checklist of measures that might be used to begin to meet this challenge, and to achieve improvements to safety strategies, safety plans and safety actions for the future.

Table 10   Managing health and safety risks

Safety Strategies	IMPROVED **leadership** and **management**
	a includes additional tests of competence for planning supervisor, designer and contractor, checking previous experience and performance
	b includes an FM audit, if employer competence is in doubt
	c considers possible use of client agent
	d encourages client involvement in design and health and safety, and promote leadership within design team if client is reluctant to take initiative.
	e promotes a positive attitude to the regulations
	IMPROVED **hazard awareness** and **risk management**
	a profiles operations, both of organisation and construction process to identify project specific risks and promote hazard awareness
	b includes user and FM insights in risk management process
	c includes contractor buildability reports in risk assessments
	d considers the use of health and safety specialist for specific risk assessments

Table 10 (Continued)

BETTER information on existing building

a   includes a full survey of existing site and building
b   checks records as hidden structure / services are exposed
c   prepares contingency arrangements when information is poor

Safety Plans

IMPROVED planning and co-ordination

a   ensures that all significant risks are responded to through:
    i   risk strategies both for 'designing out' as well as 'managing in' the identified risks
    ii operational arrangements for the elimination, and isolation of areas of compounded risk
    iii tactical measures for safe work procedures on site
b   plans early and strategically for all significant risks
c   ensures design team appointments, brief and resource, co-ordinate duties appropriate to project circumstances
d   ensures adequate time and resources for pre-planning

BETTER information on buildability

a   pre-tests construction methods wherever critical
b   uses specialist sub-contractors for hazardous operations
c   maximises use of pre-fabrication

SPECIFIC restrictions

a   include special prelims for site fire codes
b   include special prelims for site noise codes
c   include special prelims for shutdowns, isolations and retesting for all mechanical and electrical services

Safety Actions

IMPROVED communications

a   include formal exchange of health and safety policy and procedures between employer and contractor
b   promote effective on site liaison
c   support on site training in relation to specific hazards for site agents, operatives, managers and users

BETTER information on refurbishment statistics

A1 Sources of Information

A1.1 Institutes and Organisations

Association of Planning Supervisors (APS)
16 Rutland Square, Edinburgh EH1 2BE, tel 0131 221 9959

British Institue of Facilities Management (BIFM)
67 High Street, Saffron Walden, Essex CB10 1AA, tel 01799 508609

British Standards Institute (BSI)
389 Chiswick High Road, London, W4 4AL, tel 0181 996 9000

Building Employers' Confederation (BEC)
82 New Cavendish Street, London W1M 8AD, tel 0171 580 5588

Building Services Research and Information Association (BSRIA)
Old Bracknell Lane West, Bracknell, Berkshire RG12 7AH, tel 01344 426511

Chartered Institute of Building (CIOB)
Englemere, Kings Ride, Ascot, Berkshire SL5 7TB, tel 01344 23467

Construction Industry Research and Information Association (CIRIA)
6 Storey's Gate, Westminster, London SW1P 3AU, tel 0171 222 8891

Construction Industry Training Board (CITB)
Bircham Newton, Kings Lynn, Norfolk PE31 6RH, tel 01553 776677

Federation of Master Builders (FMB)
14-15 Great James Street, London WC1N 3DP, tel 0171 242 7583

Health and Safety Executive (HSE)
Construction National Industry Group, 1 Long Lane, London SE1 4PG, tel 0171 556 2100

for information see also: http://www.open.gov.uk/hse/

Health and Safety Publications
HSE Books, PO Box 1999, Sudbury, Suffolk CO10 6FS tel 01787 881165

Institution of Civil Engineers (ICE)
1-7 Great George Street, London SW1P 3AA, tel 0171 222 7722

Institution of Occupational Safety and Health (IOSH)
The Grange, Highfield Drive, Wigston, Leicester LE18 1NN, tel 0116 257 1399

Institution of Structural Engineers (ISE)
11 Upper Belgrave Street, London SW1X 8BH, tel 0171 235 4535

Royal Institute of British Architects (RIBA)
66 Portland Place, London W1N 4AD, tel 0171 580 5533

Royal Institution of Chartered Surveyors (RICS)
12 Great George Street, London SW1P 3AD, tel 0171 222 7000

A1.2 Bibliography

Barber, J (1997) Potential side-effects of the CDM Regulations. *Construction Law Journal* vol 13 no 2 pp26-37

Barrett, M (1996) Deconstructing the new construction sites legislation. *Construction Law Journal* vol 12 no 3 pp 156-172.

Bishop D (1994) The professionals' view of the Health and Safety Commission's draft Construction (Design and Management) Regulations. *Construction Management and Economics*. vol 12 pp 365-372.

Bone S (1995) *Information on site safety for designers of smaller building projects*. (HSE CRR 72/1995) UK: HMSO. (ISBN 0 7176 0777 1)

British Standards Institute (1995) *Risk management: Part 3. Guide to risk analysis of technological systems*. (BS 844: part 3) UK: BSI.

British Standards Institute (1993) *Code of practice for access and working scaffolds, and special scaffold structures in steel* . (BS 5973) UK: BSI.

British Standards Institute (1990) *Code of practice for temporarily installed suspended scaffolds and access equipment*. (BS 5974) UK: BSI.

British Standards Institute (1984) *Noise control on construction and open sites: Part 2. Guide to noise control legislation for construction and demolition, including road construction and maintenance* (BS 5228:part 2) UK: BSI.

British Standards Institute (1982) *Code of practice for demolition*. (BS 6187) UK: BSI.

British Standards Institute (1981) *Code of practice for earthworks*. (BS 6031) UK: BSI.

Brewer R, D Oleske, J Hahn and M Leibold (1990) A model for occupational injury surveillance by occupational health centres. *Journal of Occupational Medicine* vol 32 no 8 pg 698-702.

Building Employers Confederation (1993) *Fire prevention on construction sites*. UK: BEC.

Building Employers Confederation (BEC) (1993) *Risk Assessment in Construction*. London: Building Employers Confederation.

Building Employers Confederation, The Loss Prevention Council and National Contractors Group (1992) *Fire prevention on construction sites: joint code of practice on the protection from fire of construction sites and buildings undergoing renovation.* UK: BEC.

Building Maintenance Information (1991) *Safe and effective building maintenance.* UK: RICS.

Checkland P (1989) *Systems thinking, systems practice.* UK: Wiley and Sons.

Commission of the European Communities (1993) *Europe for safety and Health at Work: Safety and Health in the Construction Sector.* Luxembourg: Office for Official Publications of the European Communities.

Confederation of British Industry (CBI) (1991) *Developing a safety culture: Business for safety* UK: CBI.

Construction (Design and Management) Regulations 1994 SI 1994 No 3140 London: HMSO.

Construction Industry Research and Information Association (1998) *Implementing CDM.* CPN Workshop Report 803L. London: CIRIA

Construction Industry Research and Information Association (1998) *CDM Regulations - Practical Guidance for Planning Supervisors.* Report 173. London: CIRIA

Construction Industry Research and Information Association (1997) *CDM Regulations - Practical guidance for clients and clients' agents.* Report 172. London: CIRIA.

Construction Industry Research and Information Association (1997) *Experiences of CDM.* Report 171. London: CIRIA.

Construction Industry Research and Information Association (1997) *CDM Regulations - Work sector guidance for designers* Report 166. UK: CIRIA.

Construction Industry Research and Information Association (1997) *CDM Regulations - case study guidance for designers: an interim report.* Report 145. UK: CIRIA.

Construction Industry Research and Information Association (1996) *Control of risk: A guide to systematic management of risk from construction.* Report 125. London: CIRIA.

Construction Industry Research and Information Association (1994) *A guide to the management of building refurbishment.* Report 133. London: CIRIA.

Construction Industry Advisory Committee (1995) *Designing for Health and Safety in Construction: A guide for Designers on the Construction (Design and Management) Regulations 1994.* London: HMSO.

Construction Industry Advisory Committee (1995) *A Guide to Managing Health and Safety in Construction.* London: HMSO.

Consultancy Company (1997) *Evaluation of CDM Regulations-Fiinal report.* (HSE CRR 158). London: HSE.

Cox SJ and EF O'Sullivan (1992) *Building Regulation and safety.* BRE Report Garston: Construction Research Communications.

Cox SJ and NS Tait (1991) *Reliability Safety and Risk Management.* UK: Butterworth Heinemann Ltd.

Croner's (1995) *Management of Construction Safety* UK: Croner's Publication Ltd.

Culver C, M Marshall and C Connolly (1993) Analysis of Construction Accidents. *Professional Safety* March pg 22-27.

Davies, N and P Teasdale (1994) *The costs to the British economy. of work accidents and work-related ill health.* London: HMSO.

Dedobbeleer N and P German (1987) Safety Practices in Construction. *Journal of Occupational Medicine* vol 29 no 11 pp 863-868.

Dester W and D Blockley (1995) Safety - behaviour and culture in construction. *Engineering, Construction and Architectural Management.* vol 2 no 1 pg 17-26.

Dewis, M (1995) *Health and Safety at Work Handbook 1996.* UK: Tolley Publishing Company Ltd.

Dickson GCA (1995) *Corporate Risk Management.* London: Witherby. (ISBN 0 948691 76 X)

Dickson GCA (1987) *Risk Analysis.* London: Witherby.

Dong W, KR Sullivan and CE Rossiter (1992) A mortality pattern study on the workers in the construction industry in the UK - a pilot. *Journal of Occupational Medicine, Singapore* vol 4, no 1 January pp 31-41.

Duff AR, I Robinson, R Phillips and M Copper (1993) *Improving Safety on Construction Sites by Changing Personnel Behaviour* (HSE CRR 51/1993) London: HMSO.

Dyton, R (1996) Risk Management in *Facilities management: theory and practice.* K Alexander (ed) London: E&FN SPON.

Egbu, C (1997) Refurbishment management: challenges and opportunities. *Building Research and Information.* vol 25, no 6, pp238-347.

Egbu, C (1996) Literature review on refurbishment health and safety issues. unpublished project paper for LINK CMR 323.

Egbu, C, B Young and V Torrance (1996) Refurbishment management practices in the shipping and construction industries - lessons to be learned. *Building Research and Information* vol 24 no 6 pp 329-338.

Egbu, C (1995) Perceived degree of difficulty of management tasks in construction refurbishment work. *Building Research and Information.* vol 23, no 6, pp 340-344.

Everett, J and P Frank (1996) Costs of Accidents and Injuries to the Construction Industry. *Journal of Construction Engineering and Management.* vol 122 no 2 pp 158-164.

European Construction Institute (1995) *Total Project management of construction safety, health and environment (2nd edition)* UK: Thomas Telford Ltd.

Everly, M (1996) The Strategy of Risk *Health and Safety at Work* October , pp 18-20.

Ferguson I (1995) *Dust and noise control in the construction process.* (HSE CRR 73/1995) London: HMSO.

Fisher, J (1994) CONDAM - the new liability. *Construction Law Journal* vol 10 pp 252-253.

Glendon, A and McKenna, E (1995) *Human safety and risk management.* UK: Chapman & Hall.

Godfrey PS (1996) *Control of Risk: A guide to the Systematic Management of Risk from Construction.* London: Construction Industry Research and Information Association (CIRIA).

Grimaldi, J and R Simmonds (1989) *Safety Management.* 5th edition USA: Richard D Irvin, Inc (ISBN 0 256 06698)

Haverstock, H (1996) *The Building Design Easibrief CDM Primer.* UK: Miller Freeman plc.

Health and Safety Commission (1995) *Managing Construction for Health and Safety: Construction (Design and Management) Regulations 1994 Approved Code of Practice*. London: Health and Safety Executive.

Health and Safety Commission (1995) *Health and safety statistics 1994/95*. London: HMSO.

Health and Safety Commission (1991) *Workplace health and safety in Europe*. London: HMSO.

Health and Safety Executive (1996) *A guide to the Reporting of Injuries, Diseases and Dangerous Occurrences Regulations 1995*. (HSE L73) London: HMSO.

Health and Safety Executive (1996) *Health and safety in construction*. London: HMSO.

Health and Safety Executive (1995) *Construction (Design and Management) Regulations 1994 : The role of the client*, HSE construction information sheet no 39, London: HMSO.

Health and Safety Executive (1995) *Construction (Design and Managemen) Regulations 1994 : The role of the planning supervisor*. HSE Iconstruction information sheet no 40, London: HMSO.

Health and Safety Executive (1995) *Construction (Design and Managemen) Regulations 1994 : The role of the designer*. HSE construction information sheet no 41, London: HMSO.

Health and Safety Executive (1995) *Construction (Design and Managemen) Regulations 1994 : The pre-tender stage health and safety plan*. HSE construction information sheet no 42, London: HMSO.

Health and Safety Executive (1995) *Construction (Design and Managemen) Regulations 1994 : The health and safety plan during the construction phase*. HSE construction information sheet no 43, London: HMSO.

Health and Safety Executive (1995) *Construction (Design and Managemen) Regulations 1994 : The health and safety file*. HSE construction information sheet no 44, London: HMSO.

Health and Safety Executive (1995) *General COSHH (Control of substances hazardous to health)*. (HSE L5) London: HMSO.

Health and Safety Executive (1993) *The control of Asbestos at work. Control of asbestos at work Regulations 1987: Approved Code of Practice.* (HSE L27) London: HMSO.

Health and Safety Executive (1993) *The Costs of Accidents at Work..* London: HMSO.

Health and Safety Executive (1993) *Work with asbestos insulation, asbestos coating and asbestos insulating board. Control of asbestos at work Regulations 1987: Approved Code of Practice.* (HSE L28) London: HMSO.

Health and Safety Executive (1992) *Work equipment. Provision and use of work equipment Regualtions 1992: guidance oni the regulations.* (HSE L22) London: HMSO.

Health and Safety Executive (1992) *Manual handling. Manual handling operations Regulations 1992: guidance on the regulations.* (HSE L23) London: HMSO.

Health and Safety Executive (1992) *Personal protective equipment at work Regulations 1992: guidance on the regulations.* (HSE L25) London: HMSO.

Health and Safety Executive (1992) *Workplace health, safety and welfare. Workplace (Health, Safety and Welfare) Regulations 1992: Approved Copde of Practice* (HSE L24) London: HMSO.

Health and Safety Commission (1992) *Management of Health and safety at work Regulations 1992: Approved Code of Practice.* (HSE L21) London: HMSO.

Health and Safety Executive (1991) *Successful health and safety management.* London: HMSO.

Health and Safety Executive (1989) *Quantified Risk Assessment: Its input to decision making.* London: HMSO.

Health and Safety Executive (1989) *5 Steps to Risk Assessment.* London: HSE. IND(G)163L

Health and Safety Executive (1988) *The tolerability of risk from nuclear power stations.* London: HMSO.

Health and Safety Executive (1988) *Blackspot construction.* London: HMSO.

Heinrich, H, D Peterson and N Roos (1980) *Industrial Accident Prevention.* London: McGraw-Hill Book Company.

Hinze, J and J Gambatese (1996) Using injury statistics to develop accident prevention programs. in *Implementation of Safety and Health on Construction Sites*. Dias L and R Coble (eds) Proceedings of the First International Conference of CIB Working Commission W99 / Lisbon, Portugal. Rotterdam, Neth : Balkema. pp 117-127.

Hinze, J and Russell, D (1995) Analysis of Fatalities Recorded by OSHA. *Journal of Construction Engineering and Management.* vol 121, no 2 pp 209-215.

Hubbard, R and J Neil (1986) "Major-Minor Accident ratios in the Construction Industry." *Journal of Occupational Accidents* vol 7 pp 225-237.

Hunting K, L Nessei-Stephens, S Sanford, R Shesser and L Welch (1986) Surveillance of Construction Worker Injuries Through an Urban Emergency Department. *Journal of Occupational Medicine* vol 36, no 3 pg 356-364.

Jannadi, MO (1996) Factors affecting the safety of the construction industry. *Building Research and Information.* vol 24 no 2 pp 108-112.

Jaselskis, E, S Anderson and J Russell (1996) Strategies for Achieving Excellence in Construction Safety Performance. *Journal of Construction Engineering and Management.* vol 122 no 1 pp 61-70.

Joyce, R (1995) *The CDM Regulations Explained.* UK: Thomas Telford Services Ltd.

Kisner S and Fosbroke D (1994) Injury hazards in the construction industry. *Journal of Occupational Medicine.* vol. 36, no. 2, pp 137-143.

Kline P, R Jensen and L Sanderson (1984) Assessment of workers' compensation claims for back strains/sprains. *Journal of Occupational Medicine* vol 26, no 6, pp 443-448.

Komaki J , K Barwick and L Scott (1978) A behavioural approach to occupational safety: pinpointing and reinforcing safe performance in a food manufacturing plant. *Journal of Applied Psychology.* vol 63, no 4, pp 434-445.

Landeweerd J, I Urlings, A DeJong, F Nijhuis and L Bouter (1990) Risk taking tendency among construction workers. *Journal of Occupational Accidents.* vol 11 pp 183-196.

Laufer A (1987a) Construction safety: economics, information and management involvement. *Construction Management and Economics.* vol 5 pp 73-90.

Laufer A (1987b) Construction accident cost and management safety motivation. *Journal of Occupational Accidents* vol 8 pp 295-315.

Leopold E and S Leonard (1987) Costs of Construction Accidents to Employers. *Journal of Occupational Accidents.* vol 8 pp 273-294.

Levine D (1996) Construction contracts: clauses on health, safety and environment. *Construction Law Journal* vol 12 no 3 pp 146-155.

Levitt, R, H Parker and N Samelson (1981) *Improving construction safety performance: the user's role.* Dept of Civil Engineering, Stanford University (Technical Report 260).

Lingard H and S Rowlinson (1994) Construction site safety in Hong Kong. *Construction Management and Economics.* vol 12 pp 501-510.

Mattila M, E Rantanen and M Hyttinen (1994) The quality of the work environment, supervision and safety in building construction. *Safety Science.* vol 17 pp 257-268.

Mattila M and M Hyodynmaa (1988) "Promoting job safety in building: an experiment on the behaviour analysis approach" *Journal of Occupational Accidents* vol 9 pp 255-267.

Nanayakkara, R (1997a) *Standard Specification for the CDM Regulations health and safety file.* (BSRIA AG 9/97) UK: Building Services Research and Information Association.

Nanayakkara, R (1997b) *The CDM Regulations health and safety file.* (BSRIA AG 7/97) UK: Building Services Research and Information Association.

Nasanen M and J Saari (1987) The effects of positive feedback on housekeeping and accidents at a shipyard. *Journal of Occupational Accidents* vol 8 pp 237-250.

Niskanen T, and J Lauttalammi (1989) Accidents in material shandling at building construction sites. *Journal of Occupational Accidents.* vol 11, pp 1-17.

Niskanen T, and J Lauttalammi (1989) Accident prevention in materials handling at building construction sites. *Construction Management and Economics.* vol 7, pp 263-279.

Office of Population and Census and Health and Safety Executive (1995) *Occupational Health: decennial supplement.* F Drever (ed) London: HMSO.

Robinson, C F Stern, W Halperin, H Venable, N Petersen, T Frazier, C Burnett, N Lalich, J Salg, J Sestito and M Fingerhut (1995) Assessment of mortality in the construction industry in the United States 1984-1986. *American Journal of Industrial Medicine* vol 28, pp 49-70.

Royal Society Study Group (1992) *Risk: Analysis, Perception and Management .* London: Royal Society.

Salminen S (1995) Serious occupational accidents in the construction industry. *Construction Management and Economics* vol 13 pp 299-306.

Salminen S (1994) Risk taking and serious occupational accidents. *Journal of Occupational Health and Safety - Australia and New Zealand* vol 10, no 3, pp 267-274.

Simonds R and Y Shafai-Shafrai (1977) Factors affecting injury frequency in eleven matched pairs of companies. *Journal of Safety Research* vol 9, No 3, pp 120-127.

Smith A (1996) Health, Safety and More Prosecutions ? *Construction Law* vol 7, no 4, pp 149 - 152.

Stubbs D and A Nicholson (1979) Manual handling and back injuries in the construction industry: an investigation. *Journal of Occupational Accidents.* vol 2, pp 179-190.

Sulzer-Azaroff B (1987) The modification of occupational safety behaviour. *Journal of Occupational Accidents* vol 9 pp 177-197.

Thomas G (1996) CONDAM Regulations 1994 - Potential Difficulties. *Construction Law* vol 6 no 6 pp 217-220.

Waller J, S Payne and J Skelly (1989) Injuries to carpenters. *Journal of Occupational Medicine* vol 31, no 8, pp 687-692.

Whittington, C, A Livingston and D Lucas (1992) *Research into management, organisational and human factors in the construction industry.* London: HMSO.

Winch, G Utomari,G and Edkins, A (1997) Towards total project quality: A gap analysis approach. *Construction Management and Economics*. vol 18, no 2.

A2 Safety and health risks in construction

A2.1   HSE accident data

The HSE holds accident data on reportable injuries in accordance with the Reporting of Injuries, Diseases and Dangerous Occurrences Regulations 1995 (RIDDOR) from 1985-86 to the present.[1] The accident data used in this *Guide* is based on two sub-sets of the RIDDOR database dealing with injury severity - 'major' accidents and 'over-3-day' accidents. These data sets which cover construction industry accidents from 1985-96, are classified in three ways, by:

- **occupation**, indicating the trade of the person injured from 26 construction occupation categories
- **work process**, describing the process that a person was engaged in at the time of the accident. There are 43 process categories which consist of: construction related process; ground works, roofing, and labouring, for example; and general processes such as handling, labouring and transferring
- **kind of accident**, noting the reported kind of accident. There are 18 accident categories; fall, slip or trip, and struck by falling object for example

The two severity of injury categories are clearly defined in RIDDOR 1995. Major reportable injuries, according to Schedule 1 of RIDDOR 1995,[2] include:

- "any fracture, other than to the fingers, thumbs or toes"
- "any amputation"

---

1   The original Regulations came into force in 1985.
2   HSE (1996) *A guide to the Reporting of Injuries, Diseases and Dangerous Occurrences Regulations 1995*. London: HMSO. pp 31.

- "dislocation of the shoulder, hip, knee or spine"
- "loss of sight (whether temporary or permanent"
- "a chemical or hot metal burn to the eye or any penetrating injury to the eye"
- "any injury resulting from an electric shock or electric burn (including any electrical burn caused by arcing or arcing products) leading to unconsciousness or requiring resuscitation or admittance to hospital for more than 24 hours"
- "any other injury -"
  - "leading to hypothermia, heat-induced illness or to unconsciousness"
  - "requiring resuscitation, or"
  - "requiring admittance to hospital for more than 24 hours"
- "loss of consciousness caused by asphyxia or by exposure to a harmful substance or biological agent"
- "either of the following conditions which result from absorption of any substance by inhalation, ingestion or through the skin -"
  - "acute illness requiring medical treatment"
  - "loss of consciousness"
- "acute illness which requires medical treatment where there is reason to believe that this resulted from exposure to a biological agent or its toxins or infected material"

Over-3-day reportable injuries are defined in RIDDOR 1995 as:

- "a person at work is incapacitated for work of a kind which he might reasonably be expected to do, either under his contract of employment, or, if there is no such contract, in the normal course of his work, for more than three consecutive days (excluding the day of the accident but including any days which would not have been working days) because of an injury resulting from an accident arising out of or in connection with work" [Reg. 3(2)]

The purpose of the tables in this section is to indicate which occupations sustained injuries, while undertaking what type of construction process, resulting from what kind of accident. They are intended to provide a point of reference in terms of the relative risks of accidents for various groups undertaking construction activities.

A2.2   Major accidents in construction

There were 31,416 reported major accidents in construction from 1985-96. Tables A1 - A13 display the major accident data by frequency in various combinations to indicate historic trends.

Table A1 illustrates major accidents ranked by frequency of occurrence by process type. The four columns indicate: HSE process numbers; process types ranked by frequency; total number of reported accidents for that process type and percentage of total report accidents represented by that process type. So for example, the first row of Table A1 indicates that process number 3310 - 'finishing processes' had 7,133 reported accidents which accounted for 22.7% of all major reported accidents.

Table A1   Major accidents ranked by HSE process type.

process number	HSE process type	total number of accidents	% of total accidents
3310	finishing processes	7,133	22.70
9790	transfer on site	3,121	9.93
9980/90	other processes	2,816	8.96
3220	ground works	2,620	8.34
3320	roofing	2,300	7.32
3270	scaffolding	1,412	4.49
3230	structural erection	1,385	4.41
9780	load / unload materials	1,147	3.65
3300	bricklaying	1,128	3.59
9800	labouring on site	1,044	3.32
totals		24,106	76.71%
	major accidents total	31,416	100.00%

Tables A2 - A6 show the first five of the listed process types from Table A1 ranked by the top ten construction occupations. Over 80% of the reported accidents are represented by these ten occupations. Tables A7 - A11 show the the last five of the listed process types from Table A1, ranked by the top five occupations. These account for a minimum of 70% of the reported accidents for that process type. Each of these tables have five columns which indicate: the total number of accidents for that occupation; construction occupation; the number of accidents for that occupation and process; the percentage of the process accidents represented by the occupation; and the percentage of the total process accidents represented by the occupation. For example, Table A2 ranks the top ten occupations for the 7,133 reported 'finishing process' accidents. Reviewing the first line, a total of 3,063 accidents were reported by carpenter / joiners, of which 1,491 were 'finishing processes' accidents. These represent 48.7% of the total reported accidents for carpenter / joiners. Carpenter / joiners account for 20.9% of all 'finishing processes' accidents.

Table A2   Major accidents for 'finishing processes' ranked by construction occupation. (n = 7,133)

occupation total	occupation	'finishing' accidents	% of occ. total	% of 'fin.' total
3,063	carpenter / joiner	1,491	48.68	20.90
1,523	electrician	1,032	67.76	14.47
6,306	labourer	964	15.29	13.51
1,111	plumber/fitter	669	60.22	9.38
2,736	other construction	662	24.20	9.28
	sub-total			67.54
1,410	painter / decorator	374	26.52	5.24
443	plasterer	301	67.95	4.22
2,018	other occupations	299	14.82	4.19
1,710	all managerial	198	11.58	2.78
1,608	bricklayer	177	11.01	2.48
	sub-total			18.91

Table A3 Major accidents for 'transfer on site' ranked by construction occupation. (n = 3,121)

occupation total	occupation	'transfer' accidents	% of occ. total	% of 'tran.' total
6,306	labourer	661	10.48	21.18
1,710	all managerial	344	20.12	11.02
2,018	other occupations	321	15.91	10.29
3,063	carpenter / joiner	269	8.78	8.59
1,432	drivers	193	13.48	6.19
	sub-total			57.27
2,736	other construction	188	6.87	6.02
1,608	bricklayer	175	10.88	5.61
1,523	electrician	167	10.97	5.35
1,096	manual production	150	13.69	4.81
1,111	plumber / fitter	100	9.00	3.20
	sub-total			24.99

Table A4 Major accidents for 'other processes' ranked by construction occupation. (n=2,816)

occupation total	occupation	'other' accidents	% of occ. total	% of 'oth.' total
6,306	labourer	608	9.64	21.59
2,018	other occupations	422	20.91	14.99
2,736	other construction	252	8.71	8.95
1,096	manual production	221	20.16	7.85
1,710	all managerial	211	12.34	7.49
	sub-total			60.87
3,063	carpenter / joiner	179	5.84	6.36
1,432	drivers	149	10.41	5.29
726	maintenance	119	16.39	4.23
368	electrical distribution	95	25.82	3.37
1,523	electrician	87	5.71	3.09
	sub-total			22.34

Table A5 Major accidents for 'groundworks' ranked by construction occupation. (n = 2,620).

occupation total	occupation	'ground.' accidents	% of occ. total	% of 'grd.' total
6,306	labourer	825	13.08	31.49
895	ground worker	593	66.26	22.63
2,736	other construction	277	10.12	10.57
1,710	all managerial	170	9.94	6.49
1,432	drivers	148	10.34	5.65
	sub-total			76.83
2,018	other occupations	115	5.70	4.39
1,096	manual production	90	8.21	3.44
3,063	carpenter / joiner	66	2.15	2.52
531	paviour / roadman	58	10.92	2.21
1,111	plumber / fitter	58	5.22	2.21
	sub-total			14.77

Table A6 Major accidents for 'roofing' ranked by construction occupation. (n = 2,300)

occupation total	occupation	'roof.' accidents	% of occ. total	% of 'rf.' total
1,474	sater / roofer	1,176	79.78	51.13
6,306	labourer	295	4.68	12.83
3,063	carpenter / joiner	245	8.00	10.65
2,736	other construction	156	5.70	6.78
1,608	bricklayer	65	4.04	2.83
	sub-total			84.22
1,111	plumber / fitter	60	5.40	2.61
1,710	all managerial	52	3.04	2.26
2,018	other occupations	50	2.48	2.17
560	steel erector	29	5.18	1.26
1,410	painter / decorator	25	1.77	1.09
	sub-total			9.39

Table A7 Major accidents for 'scaffolding' ranked by construction occupation. (n = 1,412)

occupation total	occupation	'scaf.' accidents	% of occ. total	% of 'scaf.' total
1,046	scaffolder	746	71.32	52.83
6,306	labourer	213	3.38	15.08
2,736	other construction	71	2.60	5.02
3,063	carpenter / joiner	70	2.29	4.96
1,608	bricklayer	64	3.98	4.53
	total			82.42

Table A8 Major accidents in 'structural erection' ranked by construction occupation. (n = 1,385)

occupation total	occupation	'struc.' accidents	% of occ. total	% of 'stc.' total
560	steel erector	384	68.57	27.73
6,306	labourer	221	3.50	15.96
3,063	carpenter / joiner	152	4.96	10.97
2,736	other construction	139	5.08	10.03
190	steel fixer	79	41.58	5.70
	total			70.39

Table A9 Major accidents in 'load/unload on site' ranked by construction occupation. (n = 1,147)

occupation total	occupation	'load' accidents	% of occ total	% of 'ld.' total
1,432	drivers	347	24.23	30.25
6,306	labourer	309	4.90	26.94
2,736	other construction	70	2.56	6.10
1,710	all managerial	63	3.68	5.49
2,018	other occupations	61	3.02	5.32
	total			74.10

Table A10 Major accidents in 'bricklaying' ranked by construction occupations. (n = 1,128)

occupation total	occupation	'brick.' accidents	% of occ total	% of 'brk.' total
1,608	bricklayers	743	46.21	65.87
6,306	labourer	243	3.85	21.54
2,736	other construction	47	1.72	4.17
1,710	all managerial	27	1.58	2.39
895	ground worker	25	2.79	2.22
	total			96.19

Table A11 Major accidents in 'labouring on site' ranked by construction occupation. (n = 1,044)

occupation total	occupation	'labour' accidents	% of occ total	% of 'lab.' total
6,306	labourer	494	7.83	47.32
2,736	other construction	76	2.78	7.28
1,710	all managerial	65	3.80	6.23
2,018	other occupations	61	3.02	5.84
3,063	carpenter/joiner	61	1.99	5.84
	total			72.51

Table A12 shows construction occupations ranked by percentage of total major injuries. The six columns indicate: the occupation, with the percentage of total major accidents it includes in parentheses, and the three top process activities in rank order; the percentages of accidents represented by the process for that occupation; the most frequent kind of accident for the processes; percentage represented by that kind for that process; the second most frequent kind of accident for the process; and the percentage of the process that the kind of accident represents. For example, this table indicates that labourers are involved in 20.1% of all major reported accidents. That 'finishing processes' account for 15.3% of the labourer's total accidents. falls represent 45% of 'finishing processes' accidents, and injuries due to struck by falling object constituted 16.1% of the process total.

Table A12 Construction occupations ranked by percentage of total major injuries by process and injury type.

occupation	% of occ	1st kind of acc.	% of proc	2nd kind of acc.	% of proc
labourer (20.1%)					
finishing processes	15.3	falls	45.0	struck by	16.1
ground works	13.1	voltage	20.0	struck by	18.9
transfer on site	10.5	trip or slip	43.4	falls	29.0
carpenter/joiner (9.8%)					
finishing processes	48.7	falls	53.1	struck by	16.8
transfer on site	8.8	trip or slip	50.9	falls	35.3
roofing	8.0	falls	81.6	struck by	7.8
other construction (8.7%)					
finishing processes	24.2	falls	55.3	struck by	13.4
groundworks	10.1	struck by	20.0	falls	18.4
other processes	9.2	falls	33.7	struck by	18.3
other occupations (6.4%)					
other processes	20.9	trip or slip	32.0	falls	20.6
transfer on site	15.9	trip or slip	57.9	falls	25.2
finishing processes	14.8	falls	54.5	trip or slip	12.4
all managerial (5.4%)					
transfer on site	20.1	trip or slip	56.1	falls	27.0
other processes	12.3	falls	34.1	trip or slip	18.0
finishing processes	11.6	falls	51.0	trip or slip	17.2

Table A13 outlines HSE process types ranked by the percentage of total major accidents. The six columns show: process type in rank order by percentage of total frequency of accidents; the percentage of total accidents this process represents; the most common kind of accident; the percentage of accidents represented by this kind for the process type; the second most common kind of accident; and the percentage of process type accidents that this kind of accident represents. Table A13 indicates that: 'finishing processes' are the most frequent process activity type, with 22.7% of the total reported accidents. Falls account for 55.6% of the kinds of accidents for this process, with injuries due to struck by falling object accounted for 12.5%.

Table A13 HSE process type ranked by percentage of total major accidents by injury type for construction.

process	% of tot	1st kind of acc.	% of proc	2nd kind of acc.	% of proc
finishing processes	22.7	falls	55.6	struck by	12.5
transfer on site	9.9	trip or slip	48.4	falls	31.4
other processes	9.0	Falls	30.5	trip or slip	17.5
ground works	8.3	struck by	18.5	voltage	16.1
roofing	7.3	falls	79.3	trip or slip	6.4
scaffolding	4.5	falls	65.5	struck by	13.7
structural erection	4.4	falls	51.7	struck by	21.7
load/unload materials	3.7	falls	29.8	struck by	23.5
bricklaying	3.6	falls	53.2	trip or slip	14.6
labouring on site	3.3	falls	29.9	struck by	21.6

A2.3   Over-3-day accidents in construction

There are 147,868 reported over-3-day accidents in the HSE RIDDOR data set for construction from 1985-96. Tables A14 - A26 profile the over-3-day accident data by frequency in various combinations. Table A14 shows work process types ranked by frequency of accidents. This is similar to Table A1 but for over-3-day accidents. As an example the first row of Table 14 indicates that 3310 'finishing processes' were the most common process type with 39,197 reported accidents, accounting for 26.5% of all reported over-3-day injuries.

Tables A15-A19 report the first to fifth ranked processes and Tables A20-A24 report the sixth to tenth ranked processes. These repeat the format of the major injury tables.

Table A14   Over-3-day accidents ranked by HSE process type.

process number	process type	total number of accidents	% of total accidents
3310	finishing processes	39,197	26.51
9980/90	other processes	14,594	9.87
9790	transfer on site	14,543	9.84
3370	surfacing (roads)	10,310	6.97
9780	load / unload materials	9,246	6.25
3220	ground works	9,002	6.09
9840	handling	8,254	5.58
9800	labouring on site	5,951	4.02
3300	bricklaying	4,879	3.30
3320	roofing	4,445	3.01
	total	120,421	81.44
	over-3-day injury total	147,868	100.00%

Table A15   Over-3-day accidents in 'finishing processes' ranked by construction occupation. (n = 39,197)

occupation total	occupation	3310 accidents	% of occ total	% of 3310 total
19,582	carpenter / joiner	11,808	60.30	30.12
9,390	plumber / fitter	5,900	62.83	15.05
28,691	labourer	4,842	16.88	12.35
5,192	electrician	3,156	60.79	8.05
3,307	scaffolder	2,165	65.47	5.52
				71.09
2,809	plasterer	1,874	66.71	4.78
10,552	bricklayer	1,865	17.67	4.76
7,798	other construction	1,725	22.12	4.40
4,873	painter / decorator	1,440	29.55	3.67
10,954	other occupations	1,231	11.24	3.14
				20.66

Table A16 Over-3-day accidents in 'other processes' ranked by construction occupations. (n = 14,594)

occupation total	occupation	'other.' accidents	% of occ total	% of 'ot.' total
10,954	other occupations	3,868	35.31	26.50
28,691	labourer	2,020	7.04	13.84
5,765	manual workers	1,305	22.64	8.94
7,035	drivers	1,043	14.83	7.15
4,257	maintenance	994	23.35	6.81
				63.24
19,582	carpenter / joiner	914	4.67	6.26
6,183	all managerial	681	11.01	4.67
8,939	paviour	543	6.07	3.72
1,442	electrical distribution	536	37.17	3.67
9,390	plumber / fitter	499	5.31	3.42
				21.74

Table A17 Over-3-day accidents in 'transfer on site' ranked by construction occupations. (n = 14,543)

occupation total	occupation	'transf.' accidents	% of occ total	% of 'trn.' total
28,691	labourer	2,920	10.18	20.08
19,582	carpenter / joiner	1,595	8.15	10.97
10,954	other occupations	1,319	12.04	9.07
10,552	bricklayer	1,082	10.25	7.44
6,183	all managerial	1,021	16.51	7.02
				54.58
7,035	drivers	845	12.01	5.81
9,390	plumber / fitter	810	8.63	5.57
5,765	manual worker	732	12.70	5.03
5,192	electrician	616	11.86	4.24
7,798	other construction	600	7.69	4.13
				24.78

Table A18 Over-3-day accidents in 'surfacing (roads)' ranked by construction occupations. (n = 10,310)

occupation total	occupation	'surfac.' accidents	% of occ total	% of 'sf.' total
8,939	paviour	4,845	54.20	46.99
28,691	labourer	2,085	7.27	20.22
6,183	all managerial	682	11.03	6.61
7,035	drivers	674	9.58	6.54
7,798	other construction	496	6.36	4.81
				85.17
10,552	bricklayer	400	3.79	3.88
5,765	manual worker	325	5.64	3.15
10,954	other occupations	320	2.92	3.10
2,132	ground worker	132	6.19	1.28
4,257	maintenance	82	1.93	0.80
				12.21

Table A19 Over-3-day accidents in 'load/unload on site' ranked by construction occupations. (n = 9,246)

Occ Total	Occupation	9780 accidents	% of occ total	% of 9780 total
28,691	labourer	2,521	8.79	27.27
7,035	drivers	1,628	23.14	17.61
8,939	paviour	820	9.17	8.87
19,582	carpenter / joiner	677	3.46	7.32
10,552	bricklayer	642	6.08	6.94
				68.01
10,954	other occupations	524	4.78	5.67
6,183	all managerial	389	6.29	4.21
5,765	manual worker	349	6.05	3.77
9,390	plumber / fitter	299	3.18	3.23
7,798	other construction	297	3.81	3.21
				20.09

Table A20 Over-3-day accidents in 'groundworks' ranked by construction occupations. (n = 9,002)

occupation total	occupation	'ground.' accidents	% of occ total	% of 'grd.' total
28,691	labourer	2,835	9.88	31.49
2,132	groundworker	1,142	53.56	12.69
7,798	other construction	817	10.48	9.08
8,939	paviour	695	7.77	7.72
6,183	all managerial	570	9.22	6.33
				67.31

Table A21 Over-3-day accidents in 'handling' ranked by construction. (n = 8,254)

occupation total	occupation	'hand.' accidents	% of occ total	% of 'hn.' total
28,691	labourer	1,761	6.14	21.33
19,582	carpenter / joiner	954	4.87	11.56
10,552	bricklayer	684	6.48	8.29
10,954	other occupations	673	6.14	8.15
8,939	paviour	547	6.12	6.63
				55.96

Table A22 Over-3-day accidents in 'labouring on site' ranked by construction occupations. (n = 5,951)

occupation total	occupation	'labour.' accidents	% of occ total	% of 'lab.' total
28,691	labourer	2,788	9.72	46.85
19,582	carpenter / joiner	401	2.05	6.74
10,954	other occupations	357	3.26	6.00
7,798	other construction	345	4.42	5.80
7,035	drivers	281	3.99	4.72
				70.11

Table A23   Over-3-day accidents in 'bricklaying' ranked by construction occupation. (n = 4,879)

occupation total	occupation	'brick.' accidents	% of occ total	% of 'brk.' total
10,552	bricklayer	3,555	33.69	72.86
28,691	labourer	805	2.81	16.50
7,798	other construction	164	2.10	3.36
				92.72

Table A24   Over-3-day accidents in 'roofing' ranked by construction occupations. (n = 4,445)

occupation total	occupation	'roof.' accidents	% of occ total	% of 'rf.' total
3,381	slater / roofer	2,020	59.75	45.44
28,691	labourer	619	2.16	13.93
19,582	carpenter / joiner	552	2.82	12.42
10,552	bricklayer	223	2.11	5.02
7,798	other construction	213	2.27	4.79
				81.60

Table A25 which is similar to A12 in layout, indicates that labourers are involved in 19.4% of all over-3-day reported accidents. That 'finishing processes' are the most common process type accounting for 16.9% of the labourer's total accidents, with handling representing 33.3% of 'finishing processes' accidents, and, struck by falling objects constituted 22.2% of the process total.

Table A25   Construction occupations ranked by percentage of total over-3-day injuries by process and injury type.

occupation	% of occ	1st kind of acc.	% of proc	2nd kind of acc.	% of proc
labourer (19.4%)					
finish processes	16.9	handling	33.3	struck by	22.2
transfer on site	10.2	trip or slip	37.2	handling	20.2
groundworks	9.9	handling	27.3	struck by	23.2
carpenter / joiner (13.2%)					
finish processes	60.3	handling	37.2	struck by	25.2
transfer on site	8.2	trip or slip	44.0	handling	20.4
handling	4.9	handling	66.2	struck by	14.2
other occupations (7.4%)					
other processes	35.3	handling	41.2	trip or slip	18.9
transfer on site	12.0	trip or slip	44.0	falls	20.3
finish processes	11.2	handling	28.6	struck by	17.8
bricklayers (7.1%)					
bricklaying	33.7	handling	31.1	struck by	22.2
finish processes	17.7	handling	42.7	struck by	21.0
transfer on site	10.3	trip or slip	40.1	handling	22.6
plumber / fitter (6.4%)					
finish processes	62.8	handling	42.6	struck by	15.1
transfer on site	8.6	trip or slip	42.0	handling	21.0
other processes	5.3	handling	34.9	struck by	15.2

Table A26 is similar to A13 and indicates that 'finishing processes' are the most frequent process activity type with 26.5% of the total reported over-3-day accidents. Handling accounts for 34.9% of the kinds of accidents for this process while struck by falling object accounts for 20.1%.

TABLE A26 HSE process type ranked by percentage of total over-3-day accidents by injury type for construction.

HSE process type	% of tot.	1st kind of acc.	% of proc	2nd kind of acc.	% of proc
finishing processes	26.5	handling	34.9	struck by	20.1
other processes	9.9	handling	33.0	struck by	16.7
transfer on site	9.8	trip or slip	40.7	falls	18.8
surfacing (roads)	7.0	handling	49.4	struck by	17.0
load / unload materials	6.3	handling	56.3	struck by	18.1
ground works	6.1	handling	30.7	struck by	21.6
handling	5.6	handling	64.4	struck by	16.6
labouring on site	4.0	handling	37.8	struck by	19.0
bricklaying	3.3	handling	30.3	struck by	23.0
roofing	3.0	Falls	37.2	handling	22.0

A2.4   Health risk in construction

The *Occupational Health Decennial Supplement* [3] is a primary source for occupational mortality statistics in England and Wales. This report gives Proportional Mortality Ratios (PMRs) for the 17 job group categories relating to construction, as shown in Table A27. This report indicates increased PMRs relating to asbestos, asbestosis or cancer of the pleura especially among constructions workers.[4] It suggests that work with asbestos must always be carefully considered as it represents a clear health threat to all construction workers.

Table A27 summarises the PMRs for job groups related to construction. The four columns indicate: job groups for construction; a summary of the key fatal occupational hazards for each job group that can be linked with that occupation; the number of causes of death that are significantly higher than other occupational groups based on PMRs; and a summary list of examples of the significant causes of death, ranked by PMR or

3   OPC and HSE (1995) *Occupational health: decennial supplement*. F Drever (ed) London: HMSO.
4   OPC and HSE (1995) pp 127-147.

overall mortality rates . This table is only a borad summary of the information contained within the *Occupational Health Decennial Supplement* which should be consulted for more detail.

Table A27   Occupational construction job group hazards and significant proportional mortality rates (PMRs) for men in England and Wales.[5]

job group	fatal occupational hazards	no. of sign. PMRs	ranked causes of death by PMR or overall rate
managers in construction	asbestos	14	- cancer of peritoneum - cancer of pleura - cancer of eye
carpet fitters	pre-patella bursitis	1	- cancer of rectum
carpenters	machinery and falls, hardwood dust, and asbestos	25	- cancer of pleura - arteritis - falling from building
other woodworkers	wood dust and solvents	2	- bronchiectasis - melanoma of skin
electricians	electrical injury and asbestos	25	- electrical current - explosive material - cancer of pleura
plumbers and gas fitters	asbestos, lead and accidental injury	20	- asbestosis - encephalitis - cancer of pleura
steel erectors	falls and machinery	14	- falls - falling object - machinery
scaffolders	falls	11	- falls - cancer of trachea - chronic bronchitis
welders	welding fumes and gases, asbestos, uv radiation and noise	5	- pneumococcal - cancer of pleura - cancer of trachea

[5]   summarised from Appendix 4 of OPC and HSE (1995), pp 308 - 338.

Table A27 Occupational construction job group hazards and
(continued) significant proportional mortality rates (PMRs) for
men in England and Wales.

painters and decorators	organic solvents and resins, dust and falls	22	- falls - accidental poisoning - cancer of trachea
bricklayers and tilesetters	falls and cement dust	20	- falls from ladders - accidental poisoning - ischaemic heart disease
masons and stonecutters	silicaceous dust and falls	6	- silicosis - pneumoconiosis - cancer of oesophagus
plasterers	none of note	10	- cancer of trachea - chronic bronchitis - accidental poisoning
roofers and glaziers	falls and bitumen fumes	14	- falls - cancer of trachea - ischaemic heart disease
builders, etc	accidental injury and asbestos	21	- falls - ischaemic heart disease - cancer of trachea
road construction workers	asphalt fumes and electrical current	8	- motor vehicle accident - ischaemic heart disease - electric current
construction workers	asbestos and falls	29	- asbestosis - ischaemic heart disease - cancer of peritoneum

A3 Outline of CDM Regulations

Regulation 1   Citation and commencement

Regulation 2   Interpretation

Regulation 3   Application of regulations

Regulation 4   Clients and agents of clients

Regulation 5   Requirements on developer

Regulation 6   Appointments of planning supervisor and principal contractor

Regulation 7   Notification of project

Regulation 8   Competence of planning supervisor and principal contractor

Regulation 9   Provision for health and safety

Regulation 10   Start of construction phase

Regulation 11   Client to ensure information is available

Regulation 12   Client to ensure health and safety file is available for inspection

*Regulation 13   Requirements on designer

Regulation 14   Requirements on planning supervisor

Regulation 15   Requirements relating to the health and safety plan

Regulation 16   Requirements on powers of principal contractor

Regulation 17   Information and training

Regulation 18   Advice from, and views of, persons at work

Regulation 19   Requirements and prohibitions on contractors